Witness
Tree

Witness Tree

SEASONS OF CHANGE WITH
A CENTURY-OLD OAK

LYNDA V. MAPES

BLOOMSBURY

NEW YORK · LONDON · OXFORD · NEW DELHI · SYDNEY

Bloomsbury USA
An imprint of Bloomsbury Publishing Plc

1385 Broadway	50 Bedford Square
New York	London
NY 10018	WC1B 3DP
USA	UK

www.bloomsbury.com

First published 2017

Map and chapter illustration created by Jon Adams

ISBN: HB: 978-1-63286-253-2
        ePub: 978-1-63286-254-9

Library of Congress Cataloging-in-Publication Data is available.

2 4 6 8 10 9 7 5 3 1

Typeset by RefineCatch Limited, Bungay, Suffolk

Printed and bound in the U.S.A. by Berryville Graphics Inc., Berryville, Virginia

To find out more about our authors and books visit www.bloomsbury.com. Here
you will find extracts, author interviews, details of forthcoming events and the
option to sign up for our newsletters.

Bloomsbury books may be purchased for business or promotional use. For
information on bulk purchases please contact Macmillan Corporate and
Premium Sales Department at specialmarkets@macmillan.com.

For Doug, with pleasure

CONTENTS

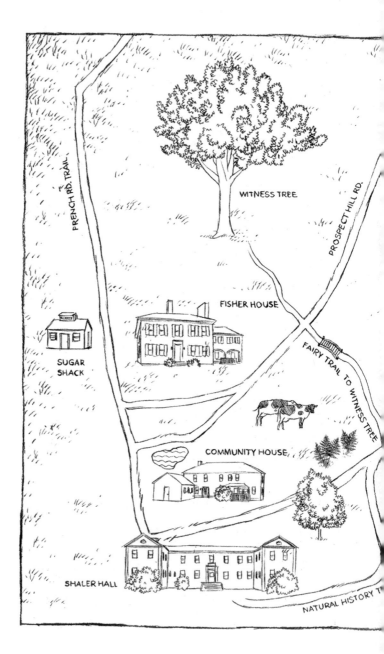

FRENCH RD. TRAIL

WITNESS TREE

PROSPECT HILL RD.

FISHER HOUSE

FAIRY TRAIL TO WITNESS TREE

SUGAR
SHACK

COMMUNITY HOUSE

SHALER HALL

NATURAL HISTORY T

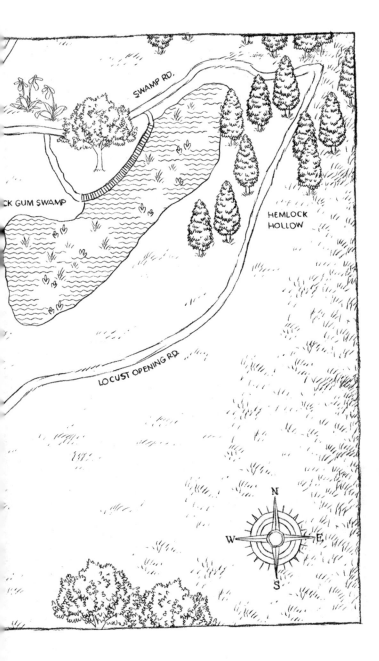

SWAMP RD.

CK GUM SWAMP

HEMLOCK HOLLOW

LOCUST OPENING RD.

N
W   E
S

INTRODUCTION

I GREW UP in a kingdom of trees. They were my constant companions in a deliciously unmanaged childhood in rural New York, lived in a ramshackle old house, parts of which date to 1763, with seven acres of woods, a falling-down barn, and a frog pond. "Go outside and find something to do" was my mother's anthem, and I took her advice. I've never really changed from the skinny kid in Girl Scout shorts of those days, my skin caked with calamine lotion from some adventure in the woods.

Grand old trees marched up the dirt lane we called the wagon road into the woods. Some had hearts of brick and arms held high with the help of pipes and wire, installed long before we got there in 1963, my two brothers, Mom, Dad, and me. There was even a spring-fed half-acre pond that shrank in dry summers to swaths of glorious stinking black mud. Bubbly expanses of algae matted the pond's warm, still waters. I can still see its deep emerald green.

But lording over it all was the four-season wonderland of our trees. In autumn, gold leaves sifted and spiraled all round our house. In the front yard, Dad defended an island of decorum, heaping leaves from the grass onto the driveway with his bamboo rake. Its scrape in the gravel as he mounded glowing piles higher and higher was the signature sound of a Saturday morning in October. When the pile reached a perfect dome about even with my seven-year-old head, it was time for me to leap into the sweet-scented chaos of crunch and rustle, feeling the enveloping leaves all around me.

No one had a leaf blower back then. Leaves, incredibly, were not for blowing and bagging, but for burning. I still remember that sweet, acrid smoke, curling from the pile in its ascent, lazy and blue to the deeper blue of the sky on an Indian summer day. Sometimes, I would get to light the pile—being a little kid, not with matches, but with a magnifying glass. Holding one leaf by its stem, I would concentrate the sun into a burning beam, watching it brown and then blacken a single hole on the leaf's surface, until it would finally light. I'd carry my flaming trophy by the stem to the pile, amazed to see how my magnifying glass had captured the sun and harnessed its energy, just as this leaf had, too.

I was the only girl in the family, with two brothers. Besides books, my other refuge was the eastern red cedar by the pond. Its bark was sweet scented, shaggy, and fragile. I could tear off sun-warmed, fragrant shreds of it with my fingernails, just to smell them—something I could never do with the stern, wrinkled frown of the maples' impenetrable cladding, or the furrowed, gnarly plating that encased the great oaks. Making my way hand over hand up the trunk of the cedar, I'd sit in its top by the hour, looking out over the pond, discovering up there the tiny immature blue-green cones that looked like berries that we couldn't see from the ground, and the peculiar pointy, stickery ends of its withes. I knew this tree's trunk well already; it was my prop for reading on the grass by the pond after school. But the suppleness of its arms, as I reached closer and closer to the top, and its sway in the wind were a surprise. Aloft, it was my first taste of a place all my own, sovereign in my tiny treetop girl nation of one.

We were all besotted with wood. On weekends, Dad disappeared to his woodshop over the garage, making tables for fun. He turned the legs on a lathe, his T-shirt blotched with wood stain, and hands holding the chisel. I'd stand across from him watching the lathe spin, enjoying its hum, and the sweet shower of shavings whizzing off the wood. We split firewood, too, in a family ritual that continued through the seasons of the year, cutting dead trees from our woodlot, my dad and brothers splitting logs with a sledgehammer and wedges. My job was holding the iron wedge for my dad, kneeling down to grasp it firm and steady, with two hands, so his blows could strike true, the wedge biting a cleft in the log. Once it was set, we'd all stand back, so he could swing big, and whack it but good. The funny little round, lifelong scar on his chest came from a hunk of iron from the thin edge of the mushroom-topped wedge Dad sent flying like shrapnel one day. I actually liked seeing that little scar as I grew to an adult, a testimony to those long-ago, shared Saturdays, and their sweet, steady seasonal rhythms. We gathered logs in the woods in the summer, when the dead trees stood out, and cut them to length with a chain saw, to split in the driveway. Come autumn, we'd heap firewood by the armload in a banged-up metal wagon Dad pulled with his Sears tractor. Rattling over the gravel-and-dirt driveway, Dad hauled our wood to the granite-slab-floored porch on the front of the house. My mom would be waiting to help us stack it, yellow jackets busy at the smell of the sap. By Thanksgiving, we'd be piling up cordwood until it covered the windows. That's what ready for winter looked like to us, from inside or out.

From October clear through May, the blazes we'd build in our fireplaces and woodstove were our household gods. Starting and tending and banking and damping—these primal tasks set the rhythms, scents, and satisfactions of our days. Our hair smelled of wood smoke; our dogs did, too. We had a woodstove for everyday use, but for big nights together, only our open fireplaces would do, brick in one room, fieldstone in another, right by the cherrywood-plank dinner table. Our dogs panted, sitting so close to the blaze our hands scorched at the touch of their fur. The sharp surprise of pants warmed by standing too close to a fire, discovered when bending to sit, is unique to people heating with wood. We had fireplace gloves of leather, and brooding iron kettles for steeping porous stones in kerosene. Lollipopped on a metal rod, we slid the stones soaked and stinking under stacked kindling, then set the whole mess alight. Good firewood, we all learned, took preparation, planning, and diligence; there was no cut corner that would deliver heat in December. Only we could do that—with help from our trees.

I was out of college by the time our parents sold that place.

More than twenty years later, at my husband's urging, I dared for the first time to go back. Arriving in soft rain unannounced in Thanksgiving week, we discovered the owners not at home and parked down the road, out of sight. I knew where I was headed: straight to the wagon road. Into the woods. Barely able to look, I walked into a moment of grace too kind for this world: my childhood trees, all still there. Still intact. Here was the old maple at the foot of the wagon road, with its heart of bricks and cement, the iron pipes and wire still sustaining its broad, spreading boughs. There, too, was the fringe of trees I remembered to the west, the branches like black lead in cathedral-glass sunsets. I found my old friends, yes, the red cedar, and the beech up in the woods I watched as a girl. There it was, with its smooth gray trunk and pyramid symmetry, light spilling through the open tracery of its branches. All just the same, just bigger and fatter. I snugged up against my beech to put my palm flat to its cool, wet bark.

Those trees had centered our world, and for me, they told time. The seasonal turn of the year made visible. I watched them from first leaf to last, then listened through winter, as they cracked in deep cold, their shadows long and blue on the snow. We wondered for years at the tumbled-down stone walls all through those woods, at the old cellar holes, and junk heaps with their crumbling tin cans. Today, I know they were left by farmers living and working on this land, part of a vast landscape of small farms and pastures once stretching clear from suburban New York into New England, long since grown back to woods.

So it was with an overwhelming sense of homecoming that I

walked into the Harvard Forest in north-central Massachusetts for the first time in the fall of 2013. Just 122 miles as the crow flies from my childhood woods, here they were once again, those same stone walls, stitching through acres of trees, pocked here and there with old cellar holes. I put my hand to the gray granite fieldstones, softened with lichen and moss. The fragrance of the leaves, the soil, the song of the bugs, the slant of the light—and the trees— the birches, maples, beeches, and oaks, hemlock and white pine. Seeing this landscape again, after years away in the Pacific Northwest, where I now live, was like savoring the pages of a family album. Observing the turn of the seasons in woods like these was something I'd done all my life, long before I knew there was a word for it: *phenology*. Or that watching how trees grow, and their leaves cycle through the seasons, had become one way scientists are studying a problem revving up in earnest since I was a child. Climate change.

A Knight fellowship in science journalism at MIT launched me on that first walk into the Harvard Forest. As a newspaper reporter, I was looking for a fresh way to understand and tell the climate-change story beyond dueling politics or science. To my great good luck, Andrew Richardson, an associate professor at Harvard University, was game when I introduced myself and my interest, graciously inviting me to spend my fellowship year visiting his lab, and joining in as he and his team of researchers did fieldwork investigating the effects of climate change on trees at the Harvard Forest. Here, at one of the world's premier research forests, in the classic New England village of Petersham,

Massachusetts, a new center of my world emerged. In this forest so like the woods I had loved as a girl, it came to me: you could tell the story of climate change—and more—through a single, beloved living thing: a tree. A specific, human-scaled focus for larger questions; a frame of study for contemplation. In the eighteenth century, surveyors used so-called witness trees as landmarks to lay out the metes and bounds of new landscapes. My modern-day witness tree would likewise be a living marker, from which to understand our past, interpret our perplexing present, and regard our future. Before long, I moved to the Harvard Forest as a Bullard Fellow in forest research through Harvard University, to find my tree and gather that story.

In the children's book *In the Forest* by Marie Hall Ets, a child picks up a hat and a horn and goes for a walk in the forest. Along the way a catenation of animals join in, each in its turn adding to the pleasure of the amble. Writing this book was a bit like that for me. Lots of people joined in as I walked the Harvard Forest and took a good long look at my tree. A big-tree hunter, a carpenter, and a forester all came calling. So did a landscape ecologist, a wetland scientist, mycologist, wildlife biologist, and all sorts of tree scientists. A historian came along, too, and an artist, and even professional tree climbers. We cored into the tree's heart, dug into its roots, climbed into its canopy, and encountered the menagerie of animals it shelters. What a busy world unfolded to us, in this one tree. We discovered it gifts a hail of acorns to birds, squirrels, and deer mice. It shelters garter snakes, ants, blackflies, ferns, and lichens. It creaks from cold in frozen nights with snow deep all around. And its bare limbs are beautifully seen in winter against distant galaxies of sharp burning stars.

Along the way, too, I saw how our carbon profligacy and climate warming are changing the timing of the seasons, and our world—and that you could even see that in just one tree. Yet I came out of the woods to write with a heart full of hope, inspired by the wonder of nature, and our place in it. I found something else, too, which I didn't expect: a new way of thinking about how we might live together into the far future on a planet we have changed in ways we never meant to.

This book is the story of those explorations, gathered mostly outdoors, on walks in every season, and from up in the tree while climbing it. Much of what I learned came from talking with scientists at the Harvard Forest and elsewhere. But I learned much, too, from the powerful act of simple observation, and being open to what one tree, and one forest, can teach. There is much in this story of weather and the forest through the seasons, its many moods, changing fragrances, the feel of it underfoot and its evanescent light, its animals and the secrets of its long dark velvety nights. For it was in the inspiration of nature and majesty of trees that I found the most wonder and joy of all.

Lynda V. Mapes
Seattle, 2016

CHAPTER ONE

ME AND MY TREE

WE WALKED IN aqueous light, the trees in their first full green leaves of summer, bright against a June-blue sky. Threading our way into the woods, where the ferns were knee-high already, we pushed through the forest understory to the grove where the big oak stood, the biggest tree in its realm. David Orwig, a senior ecologist at the Harvard Forest, rubbed a block of beeswax over the length of his metal tree corer, the better to drive it into the wood after the bite of the bit. An expert at coring trees and reading their inner secrets, he got ready to turn the slender, eighteen-inch-long corer into the tree by hand, ratcheting hard with the corer's T handle. Done correctly, the result of this type of tree sampling is a biscuit-colored, delicate core, cut by the bit in a long, slender wand. Scientists quest deep into the hearts of trees for these cores for a panoramic view into the past. Mounted, sanded, and viewed under a microscope, each band in the core of alternating color, light and dark, tells a story of the tree's life. How much rain fell, relative to other years; whether there was a forest fire or a plague of gypsy moths. The core tells of the doings of a tree's neighbors, too: whether they crowded in close or fell away, leaving a clear shot to the sun. And a core tells age, year by year. Each pair of a light and a dark ring represents a year of growth, through spring and summer.

Coring a tree is harder than its sounds, especially with an oak. It's a noisy business; trees fight back, tightening on the corer as it's pushed deep into the wood. Leave a bit in there for long and it won't come out at all; that is the grip of a tree, closing in on an intruder, and sealing its wound. Every researcher has a story of the bit he or she couldn't get out without breaking it, the metal no match for a tree on the defense.

Orwig is a master corer and a big man. I've seen him core dozens of trees in a day, and I know the sound of his work— *krrreck, kreck, kreck,* with each turn of the bit, as the tree grips the metal. The sound travels a long way through the woods and can't be confused with anything else. The rhythm of the sound is unique to each person driving the bit; every researcher plays a signature tree-coring song, know it or not. I once made a sound recording of a tree-coring foray in an old-growth stand with a dozen researchers at work. It was a garage band of trees, talking back as the scientists sought their story.

Orwig circled the oak, looking for the best point of attack. He put a hand to the oak's bark, feeling for a good spot, a bit flummoxed by the trunk's shape. Instead of rising straight, the oak leans hard away from the stone wall where it sprouted. It also grows in a twist, rising in a graceful twirl to the first branches of its vast, spreading crown, reaching some eight stories up in the air. Orwig saw a good spot. I was suddenly weirdly anxious. Does this hurt? Of course not, I corrected myself, it's a *tree*. But I had already spent a lot of time with this oak; I couldn't help feeling attached.

I first met this tree in the fall of 2013, walking these woods to study the seasonal procession of the year in the forest with John O'Keefe. A biologist given to wearing the same two sweaters all winter—that's a long time in Massachusetts—and a slouchy ragg wool hat, O'Keefe had walked the same circuit of fifty trees in this forest for more than twenty-five years. The big oak was part of his tree survey—it wears a badge of science, a shiny metal disk with its ID number in the survey: BT QURU 03. BT for Barn Tower, the location of its research plot. QURU for its species, *Quercus rubra*, or red oak, and 03, because it was the third oak added to the research plot. O'Keefe liked to say he started his long-term survey of the timing of the seasons, revealed in the budding, leaf-out, leaf color, and drop on the trees, as a way to get outside at least one day every week, then just never stopped. By now he had compiled a valuable and unique record. Seasonal changes in nature are among the most readily observable clues to the biological effects of global climate change, as warming temperatures reset the seasonal clock. In forests, water use, the growth rate of trees—their carbon storage—the length of the growing season, and temperature all are connected. So O'Keefe's work, documenting the seasonal gyre of the woods, was a look, told through the language of leaves, at our changing world.

His foot survey was the ground truth for images also beamed over the Internet of the tree canopy, continually recorded in daylight hours by surveillance cameras, watching these trees' every move, from 120 feet overhead. With O'Keefe's tree-by-tree observations, the forest-level view from the cameras, and other devices on observation towers, and even a drone used to fly regular photographic missions over the forest, these had to be among the most closely monitored trees in the world.

Scientists from around the world are studying changes in the lives of these trees, as levels of carbon dioxide mount in the atmosphere, and temperatures warm. Some trees in the Forest are martyrs, the life sucked out of them by invasive woolly adelgid, tiny bugs multiplying now that winters are too warm to kill them. Yet red oak, the dominant tree in these woods, is growing faster and more efficiently than ever recorded. You could see all this in the very breadth of the trees, in the grain of their wood, in the budding and drop of their leaves. You could, I thought, as I first heard of all this, take a deep long look at even just one tree, and see so much.

Before long, what had started out that first fall as taking a few walks with O'Keefe turned into a regular thing. Pretty soon, I had found myself renting a room in the old farmhouse next to the pasture by the woods, so I could head out into the trees whenever I wanted. But I was still looking for my one tree, to take a deep long look at its life and discover its story and, wound within it, our own. The forest I had come to visit had become a place I didn't want to leave. Why not, I thought, just move into the farmhouse and stay? "John," I wrote O'Keefe one night that fall in an e-mail, "I need a tree."

So we had gone out, on one of O'Keefe's last survey walks of the year. It was the week before Thanksgiving 2013. Frost was already in the ground, and the sun streamed through mostly bare branches. "I'm going to say that is running hard," O'Keefe said, as we crossed the bridge over a brook at the beginning of our walk. The water was flowing faster now than just the week before, O'Keefe explained, because the trees had begun to go dormant for the season. What we were seeing was an indicator that the symphony of the forest was transitioning to the slow movement of winter, the trees taking up less water and leaving more in the stream. This was new to me. What a window the streams, puddles, and vernal pools were into the lives of trees. It had never occurred to me how intimately they are connected.

Surface water, visible in stream flows and vernal pools, is depleted as trees move water from the ground to their crowns and finally breathe it as vapor into the air, through their leaves. As much as 99 percent of the water a tree takes up passes right through it, easing as water vapor through tiny pores by the millions on the undersides of their leaves. Called stomata, these microscopic mouths are the openings through which trees breathe, like us, day and night, taking in the carbon dioxide that provides the substance from which they are built. Trees are interstitial beings, connecting the atmospheric and terrestrial realms. They are rooted in the ground, but made from thin air, conjuring the sky, the atmosphere, and the sun to earthly form. For this alchemy they embody wonder; they are a transubstantiation that has amazed people for centuries. For really, who would think something so solid and long-standing as a tree could be made from the limpid, quicksilver ingredients of sun, water, and air?

We crunched over fallen leaves, releasing their fragrance. The one-note call of a blue jay found us. "That's about sixty percent down," O'Keefe said, lowering his binoculars from a maple in his survey, and making a note of the tree's percentage of leaf fall with a pencil on his clipboard. As we walked on, we were also auditioning trees: I was still on the hunt for the right one, a wizened witness to the story of our changing world. We paused long and hard at a striped maple we both admired. So lovely were its golden tresses of flowers in spring, and nut-brown seeds in the autumn. Its leaves in fall were a soft yellow not seen anywhere else in the Forest. It has snazzy bark, with thin green stripes that photosynthesize. It's an important tree for wildlife: striped maple is beloved by moose, which strip its bark and clip its tender branches for food. And it was a friendly size, with leaves and buds close enough to the ground to study up close. But it was too small; it had no age, no gravitas. And so we reluctantly walked on.

While this forest is a natural wood, reminders that it is also an outdoor laboratory and classroom were never far as we walked. Trees bristled with tags and flagging, and the forest floor was studded with equipment. There were light sensors, and laundry baskets gathering leaf litter. Often, amid the birdsong, came sounds of science, from the buzz of a drone flying a photographic mission overhead, to the hum of computers, motors, and fans. The reality is this forest is under a microscope. It's the full-time, year-round focus of a staff of about forty to forty-five biologists, modelers, GIS (geographic information system) specialists, historians, ecologists, dendrologists, paleoecologists, information and communication specialists, policy experts, atmospheric chemists, research assistants, lab technicians, and administrative staff at the Harvard Forest with an operating budget from $4.5 million to $6 million a year, and a larger cadre of visiting researchers from around the world.

On just one tract of forest, where the big oak stood, more than forty experiments were under way, involving dozens of scientists from Harvard University, and around the country and world. They were interested in everything: root respiration, deer and moose browsing habits, legacies of prior land use, and paleoecology. Forest regeneration, historical archaeology, snow modeling, and lichen population density. Acorn and maple seed production, hydrology, phenology, plants and climate change. Organic nitrogen cycling, forest history, the invasion of the hemlock woolly adelgid, plant hydraulics, and the forest-atmosphere exchange of carbon dioxide and water. Sap flow in trees, and experiments with fungi, vernal pools, and carbon dynamics. And that was just in this plot. Spread over nearly four thousand acres, the Harvard Forest, founded in 1907 and with more than a hundred years of research in the archives, has one of the longest records of some types of data anywhere, compiled in a rich, dimensional, ongoing exploration of the forest.

O'Keefe's survey walk led on into a hemlock wood, and its cool, green shade enveloped us. I could see hemlocks shredding and thinning, as the woolly adelgid destroyed them by degrees. But I didn't want to write a story of nothing but decline and loss. So we passed through their dappled light without stopping. In the black gum swamp, the light changed again, to the open sky where these deciduous trees had already lost all their leaves. I knew the oldest living dated tree in the Forest was here, a more than four-hundred-year-old prize, with bark dark and gnarled as an alligator. But black gum is an obscure species I didn't think most people would know or relate to. So we kept going. I heard a whine as we walked under the Barn Tower, one of five towers in the Forest bristling with instruments: gas analyzers, digital cameras, and more. The sound was an air pump on a device, mounted way up in the metal tower, to measure the breathing of the tree canopy around the clock.

We walked deeper into the woods and I heard something else; I was puzzled at first at the sound. That was it: a fan. Just beyond a stone wall threading through the woods, dating to the 1700s, I could make out a black chain-link fence, and beyond it, plastic-curtained, circular pens. This was Warm Ants: a multimillion-dollar, long-term experiment, with the soil inside the pens artificially warmed with blown air. The object was to see how warming the soil would affect the populations of ants freely traveling between the pens and regularly counted in pitfall traps. A propane tank, big and white as a beluga whale, hulked in the woods nearby, fueling the heating system. "Here," O'Keefe suddenly said, laying his hand on a red oak. "This might be a good one for you." I tipped my head back. We had been on walks together before, but I had never focused on this tree, off in the woods beyond the tower, on the other side of a stone wall. It was my first look at the oak.

It was big. So big, I couldn't see its top without dropping my head all the way back to my shoulders. It had no branches at all for the first forty feet, then flared to a wide crown that dominated its grove. It was a wild tree. If I wanted to climb it, I'd need professional help. Ropes, helmet, harness, the works. But it was old, that much was clear. It was beautiful, that was for sure. It was in O'Keefe's survey—I saw its tag—so I knew we would have records on its seasonal year. It was also in the view of the cameras on the Barn Tower. That put it within the full sweep of the discipline of observing these trees, from O'Keefe's boots on the ground, to using webcams and computer models to observe and analyze leaf-out, color, and drop. And there was also that stone wall.

Sprouted from an acorn there by that wall, the oak was a cultural tree, and a historical artifact. Left, rather than cut, it was a witness to all the changes that had come over this landscape. The pastures and farm fields that used to be here, delineated by that wall. The forest that grew up next, after the farmers left for jobs in the cities—jobs in industries that had both bettered and worsened our world and created the carbon emissions now torquing the seasons. A tree this big, in this spot, has seen it all, from our changing relationship with nature in our urbanized, industrialized, and digitized lives, to the altered clockwork of nature. Like the witness trees surveyors used in the eighteenth century to mark metes and bounds of new landscapes, this tree could be my marker and narrator, a living timeline of cultural and ecological change. And while it wasn't growing in a sylvan, silent setting, but rather right in the middle of science central, that was sort of perfect, too, for a look at the ways—and limits—of knowing the mysteries and wonders of trees. Now we had just needed to core it, to be sure it was old enough. Yet, when the time had come, and Orwig was about to core the big oak, my curiosity was at war with my attachment to its wild heart. The big oak wears a tag, sure, and it's in the view of the cameras. But so far, this tree hadn't been invaded by us.

Orwig pressed in with the borer. The tree did not give up quietly, squeaking as Orwig put his shoulder into it, driving in deeper. Turn by turn, he sank the bit to its hilt. Then quickly, with a light motion, pulling with his right hand and cradling the core with his left, Orwig pulled a continuous cylinder of wood from the depths of the oak, into the sunlight of day. Working with the grace of long practice, Orwig ripped the paper wrapper off a

Dunkin' Donuts drinking straw and carefully threaded the core inside. The perfect carrying case—the ongoing debate among scientists as to whether McDonald's, Starbucks, or Dunkin' straws are best is not yet over, but many is the researcher I know who works in the woods and cores trees that snags a fistful of their favorite on coffee runs. Big cores take more than one of these drinking straws, stuck together in the field end to end with masking tape. To save that bit of fuss—and the risk of wrecking a sample—some scientists special-order straws up to twenty-eight inches long from scientific supply companies. They even come in neon colors, the better to not leave them behind in the field. Straws and Ziploc bags—field science in these woods would grind to a halt without them.

Orwig shook his head and circled the tree for a better spot. "Didn't get the heart," he said, and readied for another try. A tree shouldn't be cored more than twice in a year, so this was the make-or-break sample. He took a breath, set the bit, and bore down. The second sample came out long and lovely, its wand of time a deep look into the past. O'Keefe and Orwig studied the core as it lay in Orwig's palm, doing a quick field count they would later check under a microscope. O'Keefe had guessed 100 to 110 years all along; the core showed he had nailed the age nearly to the year. At a solid century or so, it was just about right. So that settled it: the big oak would be my witness tree.

They headed back to the lab, but I wanted some time just to think, so I flopped on my back, to enjoy the dappled shade of the oak. I looked up at its canopy, the leaves fresh and new, and just listened to the crickets chirr. The sun danced way up in the top of the tree where it found the light, its leaves the brightest green they would be all year, still soft and newly emerged. As the summer solstice neared, their photosynthetic capacity was stoked by the most intense, longest sunlight of the year. For a century now, the big oak had done this ceaseless work, breathing out oxygen, cleansing the air of our pollution, all while also sheltering and enabling a whole suite of lives beyond ours: bugs, birds, lichens. Deer, mice, chipmunks, squirrels, ants, fungi, coyote, bear. I thought, too, of all the human history this tree had seen. Lying under the oak now, though, my feelings were still mixed. What were the limits, and urges, of knowing? When does a tree become a pet? A living scientific subject more an instrument than a free, wild life?

I felt something tickle and slapped my hand to my leg. A tick? The lions of the New England woods; everyone who worked here had a story about tick bites, none of them good. I whacked again at the spot and had a quick look. No. Just a fern. A soft tickle from an even softer frond. How I loved these ferns. I had watched them emerge from tight, round, coiled buds in the fructifying funk at the base of the oak in earliest May, then unfurl in a slow-motion ballet. With summer coming on now the humidity was nearly visible, and blackflies thick and biting. I regretted every inch of bare skin. I was usually decked out in my head-to-toe bugproof clothing, walking the woods looking as if dressed in pajamas. The cheapest field clothes I could find online came only in a decidedly undignified baby blue. As I turned to go, I saw sap beginning to weep from the boreholes left in the oak. I touched the sap with my finger; it tasted like water, just thicker.

Two months later, I finally came back to stay for my year with the oak, on a Bullard fellowship through Harvard University in forest research, based at the Harvard Forest. I arrived at nearly midnight over the long Labor Day weekend. For so long, I had anticipated this moment, during the wait that summer, while finishing up at work and getting ready to begin a year's leave from my newspaper job. All through the preparations and the packing, and saying good-bye to my friends and my husband across the country.

The headlights of my rental car swept over the black-dark parking lot, lighting the rain falling in silvery sheets. I hoped the key would be waiting as promised in the vestibule at Shaler Hall, the redbrick headquarters for scientists and other staff at the Forest. I pulled open the heavy, wooden door and lifted the lid on a box waiting on a side table, glad to find the envelope inside, a little smiley face drawn on the paper. Here they were, my keys to the kingdom: a snug office at Shaler Hall, and at the farmhouse an apartment I'd be renting for the year. The two-story apartment was on the north end of the farmhouse, a big, white, many-times modified grande dame. It was just a quick walk from Shaler Hall, with sweeping views to the Forest and pasture. Living here, I could walk in slippers to my tree if I wanted and explore the woods mile upon mile, every day and season of the year. I fitted the key to the lock and pushed open the door to what would be my home. I put away some groceries and headed up the wood-banistered staircase to the big bedroom upstairs and opened the windows to the rain. Its fragrance filled the room. Tree frogs were singing in the maples and oaks out my windows. Now here we were, just me and my tree, somewhere out there in the woods. I put out the lights and lay waiting for sleep, my first night of my year with the oak. I saw not a light anywhere and heard nothing but rain. The dark was total; it seemed no one was around, maybe for miles. What in the world, I wondered, had I done?

CHAPTER TWO

A BENEFICENT MONARCH

MORNING BROKE FULL of sun, the air cleared of humidity. The views of the woods and pasture I remembered, lost to me in the midnight dark, lifted my heart. I knew what I wanted to see first: the oak. I fumbled through breakfast in the unfamiliar kitchen, grateful for the coffee filters someone had left behind, since I'd forgotten them at the store the night before. It was good and hot out, still summer weather though the calendar said

September 1, 2014. In head-to-toe bug clothes, I walked up to the woods, my feet quickly finding the path to the big oak. And there it was—resplendent, verdant, just as beautiful as I remembered. I had brought pencils and notebooks, a camera, and binoculars. I figured the first order of business was to get to know this big tree, top to bottom. It would be, I soon learned, a long exploration, involving many people and all sorts of undertakings, in every kind of weather.

The big oak had grown most of its life in a forest, not an open field, and it shows in its shape. It doesn't have the wide, spreading branches of an oak with a meadow to itself all its life. An open-grown oak mounds to a round, softly shaped crown, spreading full and lush from a fat, squat trunk. It's a magnificent sight in a clear view across a broad, open space, such as a farm field. But this big oak has shed its lower branches and sprinted for the sun to beat out an encroaching forest. Its branches angle upward, joining the trunk not in the sixty-degree joints of an open-grown tree, but the sharp angles of a tree pushing skyward, competing for sun in a crowd. Its shape makes for some interesting effects. During a big rain—the kind rippling in waves driven by the wind—the big oak will do something I've not seen in another tree. A heavy rain will run down its trunk to its first rank of branches, where the twist of the tree's trunk pours the water off in a stream. I've lain underneath it, getting splattered, trying to photograph the silver cascade of rain and never quite capturing it. The water pours right into the tree's root zone, as if from a pitcher splashing into a basin. The trunk's bark in these storms darkens hour by hour, taking on a rich, glistening wet. Its lichens seem almost to cheer, and the moss on the tree's roots plumps.

It's a pretty big tree. Not a big, big tree. But decent size. Bob Leverett, cofounder and executive director of the Native Tree Society, a nonprofit association of tree scientists and aficionados, came out to the Harvard Forest to measure the oak with me not long after I moved in. A former air force engineer specializing in computer science and statistics before retirement, Leverett is a serious tree geek, and a committed big-tree hunter. He arrived packing thousands of dollars' worth of equipment with unpronounceable names he kept pulling from his pockets and the depths of his gear bag. Lasers and counters, a tripod, measuring tapes, and more. Leverett talked to himself as he circled the tree, eyeballed it, and walked back and back with his eyes on its crown, trying to figure out just where its top even was, not an easy thing with a forest-grown tree. "Parallax effect, got to compensate for that," he muttered, looking through the viewfinder of one of his gadgets. "Nothing is making sense." Leverett called out his measurements, while I wrote them down. "I'm discounting the hundredth places on the decimal," he explained, and studied the tree from a new angle that seemed to somehow be more correct than the first. "Now I am happy. Those are reasonable numbers," Leverett said as I scrubbed out the first try with an eraser. Clearly, he had an obsession. Nothing unusual about that; for some people it's chess or African violets or saving the sea turtles . . . but why measuring trees?

For Leverett, it's about telling the story of a lost world. With so many big trees picked off in the waves of cutting since the first arrival of Europeans, most people today don't know, outside of a few remnants, what a really big, old tree even looks like. "It's like turning eagles into chickens, we have robbed a species of its dignity," Leverett said. So he hunts the woods for the old guard, scouring forests all over in his home state of Massachusetts and beyond for the giants that remain. He wants people to appreciate what a tree can do, if left alone long enough in one place—several hundred years at least. It's not only the nobility of a really big, old tree, but its power to define the ecology of a place, in everything from the soils and water and atmospheric exchange, to temperature regulation, and the animals it shelters and feeds. It's also a personal thing with him and big trees. "The real reason is in here"—he tapped a finger over his heart. All his life, trees had been a lifeline for Leverett, and the bigger, the better. "Whenever I was under a lot of pressure, I would find the only respite I got was to go into the forest," Leverett said. "I need trees for my emotional stability and health."

A red oak in these sorts of woods can live to three hundred or more years. The current reigning state champion is a 326-year-old beauty David Orwig hunted down in 2006, in a state reserve at Wachusett Mountain. The big oak we were measuring at the Harvard Forest was nothing like that. It is in truth more of a middling oak. After more than an hour of field measurements, consideration, and then more calculations later, back home at his desk, Leverett worked up his best guess of the oak's vitals. Like Orwig, and John O'Keefe, he estimated the tree's age at about a century, perhaps a bit more, maybe 115 to 118 years. He measured its maximum height at the top of its crown at 83.5 feet, and its circumference at the base at 10 feet. Measured at five feet up its trunk, the tree's circumference was 8.2 feet, and its diameter 32 inches. Its average crown spread was a big, generous 60 feet. It's a good, medium-size oak, and about right for its age—perhaps a little big. But then, the oak is growing like a teenager, faster than at any other time scientists here can document. But I'll get to that.

Before I set out to measure the tree with Leverett, Audrey Barker Plotkin, site and research manager at the Harvard Forest and a professional forester, came out to look at the tree, too, not only with some of the same tools that Leverett uses, but something different: a forester's eye. As I was getting to know the tree, I embarked on a series of what I called tree soirees, bringing different sorts of people out to meet the tree and to see what they made of it. As with an art piece, I knew people would bring their own life experiences and orientations to what they saw in the tree. I brought the perspective of a storyteller looking for a main character and a story line. What would others see? I was at the beginning of taking the tree's measure and wanted to assess it from as many angles as possible. A carpenter, a mycologist, an urban planner, a professional tree climber. A farmer. An artist. And a forester: Audrey Barker Plotkin.

Like Leverett, she triangulated, taped, and measured. But she also thought about the tree in terms of its usable wood. She saw about three decent logs in the tree, plus some firewood, maybe 1,304 board feet in timber in all. It would take about twenty trees its size to frame the average twenty-four-hundred-square-foot North American home. It had no veneer-quality logs, not with that twist. Red oaks like this one are just good, solid utility players for many farmers and other people making a living from the land out here, said John Wisnewski, the head of the Harvard Forest Woods Crew.

I encountered Wisnewski for the first time up at the Barn Tower on a crisp fall morning, getting a tool from the maintenance shed. Even in a fancy Ivy League forest with millions of dollars of scientific equipment in the woods, somebody still has to plow the snow, cut and haul the firewood, deal with the pipes when they freeze, or climb up into a tree to fetch a drone out of the branches when some scientist's gadget has gone missing. The Harvard Forest Woods Crew embodied a lot of the field station spirit and sinew of the place and held a trove of its local and institutional knowledge. There wasn't anything they wouldn't or couldn't do, it seemed, from clearing brush to knocking down a house wall for a remodeling or rounding up an escaped cow blocking traffic on the highway. The five-man full-time crew during my time there was headed by Wisnewski, a tough-tender type whose lineage in these Massachusetts hill towns went back generations.

We talked a bit about red oak as a species. He sees them as a long-lived, no-fuss tree a family can turn to when it's time for some cash. "It's the tuition tree," Wisnewski said, and many also are the taxes that have been paid with a harvest of red oak. It was long a lumber of choice for building, too, after chestnut blight beginning in the 1900s took out the big American chestnut trees.

Most of the old growth white pines were already cut. That pretty much left red oak for the heavy stuff in construction. That took a personal turn for me not long after Wisnewski and I talked.

On one of those perfect fall mornings, as the summer humidity relented to days cool enough for a corduroy shirt, I was startled, making breakfast, by the sound of a ruckus out back. I went outside to discover a construction crew arrived to *jack up my house.* It seemed that the old farmhouse needed a new sill. The original red oak beams, resting on granite blocks, were succumbing to dry rot. And suddenly, here came Wisnewski, his John Deere tractor rumbling into the backyard, the front-end loader piled with new sills. They were solid red oak, cut that summer by the Woods Crew from blowdown harvested after a summer storm and sawn into eight-by-eights at the Forest's own sawmill right across the pasture from the farmhouse. And now here they were, the fresh-cut wood fragrant, with a bit of lichen and bark still clinging to spots where the saw left some of the logs' raw beauty. I ran my hand over the beams, marveling at their size. My palm could not cover the ends.

Over the next several weeks, two carpenters arrived each morning to work at installing jacks to inch up the old house and ease the big new beams into place. They had to notch them first, hand-cutting lap joints in the ends with a hammer and chisel. "I love working on old houses," said Joe Horstman Sr., who arrived with his cohort Brian Wilkman each morning in a white box truck. Two guys from postcard-pretty rural Massachusetts, they were quite the pair, Wilkman in his camo T-shirt, and Horstman with the spare good looks of an old New England sea captain, white beard and all. We got to know each other well, what with their daily seven A.M. arrivals while I was still making my first cup of coffee. Wilkman won me over by bringing along jars of jam he cooked up himself from the wild Concord grapes just then coming into perfect ripeness all over the woods, festooning the trees in fragrant purple cascades. That jam was the essence of grapeness, and ecstasy on peanut butter breakfast toast.

Midway through their work on the house, on a crummy rainy, cold day, I saw Horstman eating lunch alone in his truck. Wilkman hadn't shown up; he was AWOL, most likely turkey hunting. I coaxed Horstman out of his truck with an invitation to come visit the big oak. I was curious to see what a carpenter would find in it that I didn't. Horstman lives in his parents' old house, built more than a hundred years ago, a post-and-beam beauty crafted from chestnut and oak. He sees trees as individuals. "There are so many, they seem alike, but each is different, it has its twists and turns, you can walk by a million trees and never see the same thing," Horstman said. It's the same to him with lumber; no two pieces are the same. "I like the smell from all the different woods. When I cut a piece of wood, I have to smell it, and I associate woods with its smell. Oak is fruity, it has that tang." He took a few chunks of scrap oak home from the sill project, Horstman said, and put them in a bag with his hunting clothes, to give them the fragrance of the woods. "You wonder," Horstman said, looking up at the big oak, "why did it grow in that twist? What has it seen, all these years, walking underneath? You think of all the storms, the snows, the ice, it has been through, and all the people who have walked beneath it. And the animals. And it still has a lot of life left."

One day I joined Horstman and Wilkman in the yard as they worked at the beams with hammer and chisel, the sweet-smelling oak curls piling up on the grass. "Here, let me show you something," Horstman said, putting down his tools to root around in the back of his truck. He came back with a hunk of wood silvered and gnarled with time: a piece of the old beam from under the house. Today it's a bookend on a shelf for special things in my study, a look back in time at the truly big trees that Leverett talks about, its rings tightly nested by the hundreds. The big new beams I had been so impressed with were nothing compared to this; their grain seemed wide and loose by comparison, and their density was nowhere in the same class. Just as Leverett had said, we forgot—if we ever knew—what big or old really looks like. All the same, I liked knowing that the house I'd be living in would be resting on red oak sills, cut from these very woods, beams that would hold this house steady and true for another century at least. "Keep it dry, and it will just last and last," Horstman said of red oak.

Oak has always been important to people. It's not rarity or anything unique about oak that makes it special; many trees are bigger, stronger, or more exotic, with specialized niches in challenging terrain. But with oak, its ubiquity and broad utility are exactly what makes it practically family. If oak were an animal, it would be a dog. The most common tree in the northern hemisphere, with some six hundred species in all, oak grows in every situation, from lowlands to rocky scree, from Greenland's icy mountains to the coral sands of India. The first oak fossils from 45 million years ago show remains of oak dating back to the Eocene Epoch, long before people. We put oaks to use from the start. The acorns fed not only cattle and hogs, but people all over the world. *Oakcorn* is the origin of the word *acorn*, and people since the earliest times, including Indian tribes in North America, ground acorns into flour for bread.

Norsemen cruised the seas in their Viking ships hewn from oak. The British Royal Navy's ships were built with white oak, from forests protected by the Crown. Ancient Britons held the oak sacred, and their priests, the druids, used groves of oaks for their temples. Many of the oldest wooden buildings in England today were constructed of oak. The English countryside is dotted with venerated old oaks, named and known celebrities of the vegetable kingdom. And why not? The ancient Yule log of the druids was always of oak, and King Arthur's Round Table was made of oak. More places are named for oak than any other tree; the Aclands, Actons, Oakleys, etc., all take their name from oak. Air-dried English oak weighs in at forty-eight to fifty-five pounds per cubic foot, and the oldest trees easily range four to eight feet in diameter and are up to six hundred or even a thousand years in age. At least one English king took shelter in an oak, outwitting enemies in pursuit.

In the United States, presidents privileged to plant a tree on the White House lawn chose oak more often than any other species. Franklin Roosevelt and Lyndon Johnson may not have had much in common with J. Edgar Hoover, but they all planted oaks. The Charter Oak at Hartford, Connecticut, destroyed by a storm in 1856, was a landmark of early colonial history. Its picture is on the three-cent stamp issued in 1935 for the Connecticut tercentenary and abides today on Connecticut's quarter. The name comes from local legend in which a hollow of the tree was used to hide the colony's self-governing charter from British officials seeking to return it to the Crown and restrain the colonists' independence. Indians were said to have held council under its boughs. When the mighty white oak fell, the people of Connecticut had pieces of it made into a ceremonial chair adorned with carved acorns and leaves of oak. The chair is still in use today, on the rostrum of the state Senate.

In his book *Republic of Shade*, Thomas Campanella notes that trees are totemic objects in American civic and cultural life, celebrated for their age, especially in such a young nation. Lacking cathedrals and ruins and crumbling castles, our relics were from the natural world—old trees, mountains, and venerable rocks. "Like the marble scattered over the Appian Way, the aged trees served as a yardstick of time," Campanella writes. Repositories of grace and gravitas, big trees anchored civic and home life, their very rootedness an aspiration for the people dwelling under their leafy bower. That's just as true today, perhaps even more so in our evanescent digital age. Trees are time made visible. They offer the assurance of a presence enduring beyond our own, a cheat to our fleeting mortality.

Members of the family Fagaceae, oaks are part of a large,

ancient group of flowering plants, including beeches, hazelnuts, and many fruit trees. They grow in bewildering variety. Some species are so dense they sink in water, while pieces of *Quercus suber*, its bark peeled to make the corks for wine bottles, will float. Oaks have descended to six present genera of *Quercus*, the name derived either from the Celtic *quer cuez*, or "fine tree," or the Greek *choiros*, or "pig," an animal that loves acorns. I've seen both etymologies and like them equally. Plants in the family all bear catkins—the long tassels of tiny flowers whose windblown pollen is a torment to many in spring—and fruits enclosed in a husk called an involucre, or cupule—the cup of an acorn.

I think about the big oak's origins at the Harvard Forest. Did it germinate from an acorn left behind by a squirrel or perhaps dropped by a bird? Maybe a chipmunk tucked it away in the stone wall it grows next to and forgot to come back for its meal. It's a wonder it sprouted at all. A red oak won't begin to flower for at least twenty-five years, and it takes fifty years to put on a good size crop of acorns. I don't know where the big oak's parent tree is, but I doubt it is nearby. The big oak is the largest oak in sight from its trunk.

Acorns take two years to grow and ripen, forming from the female flower in early spring, falling to the ground from August to October, and chilling over the winter. The acorns of a red oak will not sprout until they have been through at least three months of temperatures below forty degrees Fahrenheit. But then, big and meaty, and packed with carbohydrates, fat, and protein, they are ready to go. Each is a dynamo, encasing the nutrition needed to launch a mighty oak, and all of the essential parts of the infant tree, perfect and complete: roots, stem, leaves, and buds. In early spring I saw acorns germinate as soon as the snow melted out on the old roads winding through the Forest. First, they swelled with the spring rains. Then they split their smooth brown shells, showing the meaty, living pink seed inside, cleft in two. A bit of root reached out: the taproot drilling into the ground. If it doesn't dry out—and most acorns on top of the ground will—the taproot will screw deep into the soil, anchoring the young sprout. Acorns buried under leaf litter do better—and not only because of moisture the leaf litter conserves. One of the chief hazards for an acorn is being eaten long before it can sprout.

In fall, sitting under the big oak can be a percussive experience, as acorns sail to the ground with a resounding thwack. As soon as the crop is set and ripe, birds, especially jays, will raucously harvest, working the tree's branches rather than losing out in the race for acorns on the forest floor. The wood becomes visibly more busy. The stone wall by the tree becomes a highway for chipmunks; squirrels come in from the four directions, and blue jays fly missions into the woods from all over. Sitting under the tree, I once watched a chipmunk stand its ground on the stone wall, startled to see me, but then refusing to leave, despite my alarming and annoying presence. Tiny and determined, there it was, waiting for something to happen that would clear its way to get to the round hole at the base of the tree's trunk, where, I presume, it either had a den or a cache, or both. I watched this hole for the chipmunk's activity and enjoyed seeing the bits of shell and acorn detritus scattered about from foraging missions. What a provider this tree was. But the benefits were not only one-sided.

Oaks are shade intolerant, and their seedlings fare poorly under the parent tree. To make a go of it, somehow the acorn has to germinate a good distance away. Enter the delicate balance, says biologist Mike Steele, between the oak, the blue jay, and the gray squirrel. Steele, associate professor of biology at Wilkes University in Wilkes-Barre, Pennsylvania, studies the ecological and evolutionary interactions between jays, squirrels and other rodents that disperse seeds and acorns. "I became very interested in what animals do with the seed, and the results of that behavior in the establishment and regeneration [of trees]," Steele said. The oak needs the acorn to be attractive enough to the jays or squirrels that they will pick it up and carry it to a cache—but they can't eat so many that there will be none left to create a new generation of oaks.

Many a tree owes its life to the forgetfulness of squirrels. But their memory is still mighty good. Gray squirrels, Steele says, triangulate the location of a cache based on stationary objects. Putting the squirrels in captivity to test their memory, Steele discovered that even three weeks later they would immediately find their food stash. Consider that the next time you walk into a room to do or get something and can't remember once you arrive why in the world you went in there.

Squirrels are quite discriminating, moving only the largest, choicest acorns farthest from their source—the smaller ones aren't worth the effort. That's also good for the tree, with the most viable seeds winding up farthest from the parent. As so-called scatter hoarders, squirrels also create many caches, not a single trove—another plus for surefire dispersal. The animals even risk a higher predation threat to bury a choice find out in the open— perfect for a tree that needs sun. They work hard at their task, regularly checking on and moving their caches to safeguard them. They will even deliberately mislead competitors, pretending to hide an acorn, pushing it into the ground. But an animal that goes to the spot to dig it up will find . . . nothing. "What we found suggests that squirrels are engaging in functional deceptions," Steele said. "In any videotape, you would swear they are going through the same routine" of burying a cache.

Squirrels are also extremely sensitive to insect-infested acorns. A squirrel will do a rapid head shake with an acorn to assess the seed quality before bothering to take it. Researchers determined through X-ray examination of the acorns that the squirrels were identifying weevil-infested nuts with about 92 percent accuracy. "They selectively cache the sound ones," Steele said. Apparently once the weevil feeds on the cotyledons within the acorn, it is no longer the same nice tightly packed package of nutrition. The animal can detect the looseness with a head shake that is so rapid, researchers had to slow down their videotape playback to detect it. "There is one surprise after another," Steele said. "I couldn't have sat around and made this up." Next he is interested in researching how the landscape of animals' fear—how far they are willing to venture from their nests and burrows to find and stash food—influences the dispersal and growth patterns of forests. Theirs, after all, is no minor impact: Steele says he has watched animals make eight thousand acorns disappear in forty-eight hours.

I once watched this dispersal business in action myself on a sunny autumn afternoon. As I pegged out my wash on the clothesline beyond my house, my eye was caught by a repetitive motion in the pasture just beyond the board fence at the back of the house. It was the regular lazy arc of a blue jay, flying from the forest, some half mile distant, across the pasture to big old oak trees that line the dirt road up into the woods. It was a single jay, a big brilliant-blue male shining sapphire in the sun. He'd busy himself in the oaks just a few moments, then fly with an entirely different pattern—a fast beeline, full of purpose—straight back to his cache in the woods. A few moments later, he was back again, foraging in the same trees. He was doing his work, and I was doing mine, both of us making the most of a fine autumn afternoon.

The life history of oaks and the animals that depend on them are so wound together, it affects everything from the pattern of where trees grow in the forest, to the timing of acorn production. Oaks will mast—produce large crops of acorns—only once every two to five years. The trees' reticence is a strategy that helps outsmart the many animals—more than a hundred vertebrate species in the U.S. alone—that feast on acorns: flying squirrels, raccoons, deer, chipmunks, squirrels, voles, black bear, mice—and all sorts of birds. Blue jays are the long-distance dispersers of acorns, taking them two and even three miles from the parent plant. Woodpeckers, tufted titmice, grackles, white-breasted nuthatches, sapsuckers, quail, and ruffled grouse all feast on acorns. A wild turkey will eat more than two hundred acorns in a sitting. In the fall, deer will make up more than half of their diet with acorns, available just when animals need them most because other sources of food are depleted. So heavily targeted are acorns as food that only one out of five hundred will produce a one-year-old seedling. But perhaps the odds used to be worse. Passenger pigeons, once the most numerous birds in North America—their numbers equivalent to that of all of the bird species overwintering in the United States—were huge consumers of acorns. Today the great flocks are no more.

Evolving amid such heavy animal use of acorns, oaks adopted a strategy of setting out a big banquet only every couple of years. That keeps the animal population that dines on acorns proportioned to the smaller crops. Then oaks overwhelm the locals' appetites every so often with a bumper crop, putting out many more acorns than can possibly be consumed, ensuring some will be left to seed a new generation. These mast years are a fiesta in the forest with wide-reaching effects on wildlife. Deer have more twins. Other animals, including mice and black bears, show a population boom. On the downside, the tick population explodes, too, with all those nice warm-blooded animals to feed on. Such is the way of the forest, where everything is wrapped into the life and the cycle of everything else.

I think of the big oak not only as a dominator of its realm, crowding out other trees from the sweep of its vast canopy, but as a beneficent monarch, enabling a broad suite of lives. It provides not only food with both its acorns and leaves, but shelter, too, especially in fall with its leaves lasting late into the season. What a menagerie lives in the airy castle of a big oak and frequents its kingdom of shade. Red efts, delicate salamanders no longer than a pinkie, cruise the moist, cool duff of leaf litter. Their smooth orange skins glow with bright yellow dots, and their delicate toes grip the fallen leaves with aplomb. Tree frogs sing in the boughs, and owls nest, gaining a good high view of their next meal. I've watched delicate spiders make their living combing the gnarled wilderness of the big oak's bark for tiny bugs. Lichen splotches its trunk like splats of thrown willow-green paint. On its north side, moss grows thick on the oak's humped roots. Smooth-skinned garter snakes slink through its shady realm to bask in a sun fleck, and a bird nation of foragers and songsters are drawn to the insects it hosts. Even in winter snow, I've noticed the deep clefts of the big oak's roots sheltering green shoots of fern.

One wild creature can kill an oak: the gypsy moth caterpillar. In a severe outbreak, gypsy moth larvae can strip a tree in two successive seasons, and in a dry, poor site, that can kill a tree. In the big oak's core, telling its life history, more than one tight, tiny band tells of tough years: 1944 and 1981, two years in which records in the Harvard Forest Archives document ravenous gypsy moth attacks. While years such as those will make headlines and history as they kill and stunt trees, every year caterpillars have at least some impact. But here, too, the oak has evolved a defense.

An oak under heavy caterpillar assault in spring can change the chemistry of its leaves so they become unpalatable and even emit an attractant in the air to call in a defensive air force of predators to devour its tormentors. Neighboring oaks can "eavesdrop" on the struggle, detecting the chemicals emitted by the oak, and gear up for battle, too, even before being attacked.

Nonetheless, by season's end it seems nearly every leaf dropped to ground shows a bite, tear, or hunk taken out. Some leaves even wear the baubles of gall wasps, a round ball or lump the insect will induce the tree to construct. Irritated by eggs laid by the wasp, the tree will isolate the offended tissue by walling it off—thereby inadvertently constructing a snug home for the developing eggs. The emerging larvae drill a pinhole in the firm round gall. I mostly find these galls on the forest floor in autumn, fallen to the ground after the whole bizarre cycle has run its course. These galls are often colorful and marbled, like Venetian glass. I have found them in all sizes, from tiny as a pinkie nail to big and fat as a walnut. But inevitably, when I bring them inside to admire, they dry to brown and shrivel like a raisin. My desk at Shaler Hall was always covered with leaves, twigs, bark, and acorns from the big oak tree and my walks in the forest. Especially with my office door shut, I was steeped in the scent of the forest, emanating from my desktop natural history museum.

Oaks are both male and female at once, and in the spring the trees are festooned with dangling bright green catkins that puff out the yellow-green pollen the wind carries to tiny female flowers, small as the point of a pen. Each spring, puddles and vernal pools in the forest and even the cars in the parking lot at Shaler Hall were dusted with pollen, a gold-green dusting I saw at no other time of year. It was the color of new life.

I watched the big oak for its first leaves all through March and well into the third week of April, as a cold, late spring followed what turned out to be a record cold, snowy winter during my year at the Forest. (Just the sort of interannual variability that so confuses the conversation about climate change, but I will get to that later.) But May suddenly turned and stayed record hot, with temperatures even into the upper eighties. Then sure enough, one bright, breezy afternoon, as I headed into the forest, I looked over to where my tree grows and saw a new complexity in the skyline of trees, a softness in what had been the spare geometry of winter. Leaves. Well, not yet leaves, but more like a yearning for them, if trees can be said to yearn.

I went to the big oak and lay underneath it with my binoculars and studied its branches. Sure enough, the swollen buds had finally begun to break open. Bright, tender new leaves, no longer than the first joint of my smallest finger, were lit by the sun at the top of the canopy, a green I would not again see all year. Soon, their color would change, taking on a rosy glow from the sunscreenlike protectant the leaves develop—in addition to a copious downy white hair. When the leaves are natal and new, their fuzz also provides protection from drying wind and burning sun. It can take as long as six weeks for oak leaves to grow out and stiffen, gradually acquiring the deep olive-green color and waxy cuticle on their tops that will equip the leaves for another summer's work. While from a distance the leaves look the same all over the tree, they are anything but.

The late Steven Vogel, professor emeritus of biology at Duke University, devoted a big part of his career seeking to understand the wonders of leaves. He wasn't a botanist, but a scientist who wanted to examine the physics of everyday life, using the leaf as his subject. This all started for him, as it so often seems to in science, with fruit flies. In the 1960s Vogel built a low-speed wind tunnel to study fruit fly flight mechanics, but quickly grew tired of that. "They got me my Ph.D., but trying to fly fruit flies in a wind tunnel with little wires on them is a way of getting old fast," Vogel said. It was time for a new line of inquiry. "I was casting about for a system that was a little more tractable, that would be fun experimentally, and I thought, well, what about a leaf?" A lot of wonderful physics is involved in a leaf's doing what it has to do—getting water, making food—Vogel learned: "Making a big broad leaf like you have on an oak, it's a pretty remarkable thing."

His interest in the physics of flow morphed to understanding how leaves behave in wind, a subject he would return to again and again over fifty years of inquiry, including the question of how leaves avoid overheating in hot sun when air movement drops to almost zero. This was low-speed fluid mechanics research, Vogel said, "without the really hair-raising complication of flapping flight."

Oaks were the first trees he worked with to explore sun- and shade-leaf differences. The variation of leaf shapes on a single tree, he discovered, was the key to their heat tolerance. A leaf on the south side of the tree or the top has to worry about overheating. The small sun leaves on the big oak are an adaptation to deal with leaf temperatures that get above one hundred degrees. "You can make yogurt at one hundred and twenty degrees," Vogel said. "You are at the upper edge of what proteins will do." A smaller leaf blade, with deep scalloping, allows no part of the leaf to be far from the edge, allowing for better heat transfer. Vogel said, "You want a lot of edge, and not a lot of blade." Meanwhile, a leaf not in the sun can afford to have a nice wide blade. Shade leaves are broad and luxuriant so as to gather what light they can. Their area is many times that of sun leaves, so much so, they can look as if they are from another tree altogether. Putting leaves in a wind tunnel opened more discoveries for Vogel about the mechanics of leaves: "They did all kinds of wonderful things. They rolled up into cones, they curled, and clustered. No one had ever looked at what a leaf does in a potentially destructive wind in terms of the mechanics."

The biggest mysteries though are underground. "The soil is the last frontier," said Serita Frey, professor of soil microbial ecology at the University of New Hampshire, and one of many Harvard Forest collaborators with long-term experiments under way at the forest. "Leonardo da Vinci is quoted as saying that 'we know more about the movement of celestial bodies than about the soil underfoot.' And that statement is still true five hundred years later." In her work, Frey has paid particular attention to the microbial life in the soil of mixed hardwood forests.

It often seems still and quiet out in the woods—but it is

anything but, even in the soil. I've wiggled my finger into the ground under the big oak and found it to be so much more than dirt. It is riddled with fibrous roots right at the surface. The earth below the leaf litter is sweet-smelling, rich dark humus, full of life seen and unseen, with more than a thousand living root tips in a teaspoon of soil. Lying on my side on the forest floor under the tree, just to watch life go by in the litter layer and a tiny soil pit dug with my fingers, I've never had a dull moment. Tiny creatures slither and scuttle in busy lives. Bits of material are shoved here and there. Creatures small enough that I squint to see them are, I know, just a hint of what's down there. "I tell my students if you want to do something unique, be a real explorer, the soil is where it is at," Frey said. "It's crowded down there, it's a really complex physical and chemical matrix. I visualize it as a cabin with large and small rooms, with large tunnels that connect them. You could walk through some, and there are small ones you'd have to crawl through on your stomach, and some you couldn't get through. And then you have these organisms that span a scale much larger than the smallest things we can see, to the largest. Insects up to elephants and whales—that is our range. For bacterium, I liken that to the piece of paper that comes out of a hole punch, and I scale up from there. An earthworm is six football fields long, it basically stretches across campus." Even at the hole-punch level and smaller, the diversity is dazzling, with billions of bacterial and fungal lives, each suited to its own niche. "They may be doing something entirely different from another colony of bacteria that are just microns away."

It is easy to forget the extensive root network also thriving below my tree, and how far it continues out from its trunk. Typically, tree roots extend beyond the trunk one and a half times the height of a tree. The roots aren't deep; mostly, they are in the top foot of soil. But the tree's life underground doesn't stop there. A fantastic unseen network of fungal filaments also intertwines with the big oak's roots and the soil, carrying on a constant give-and-take with the tree. Usually visible aboveground only as mushrooms, the fruiting bodies of these fungi are just a hint of the actual fungal organisms underground, which often are much larger, spreading in busy networks of filaments. The thin threads of these mycorrhizal fungi grow in and around the tree's roots, insinuating themselves into tiny interstices, living out their days in a symbiosis between the tree and the fungus. They typically are just five to ten microns in diameter—less than one twentieth of the width of a human hair. Small, but mighty.

The tree provides food it makes using the sun to the fungus, needy for sugar in its kingdom of dark belowground. The fungus benefits the tree by lounging all throughout the soil matrix. This second root system, if you will, greatly increases the actual roots' surface area. That aids the big oak in the uptake of water and nutrients, including nitrogen and phosphorous. The volume of these fungal filaments, or hyphae, is enormous. "If you had a microscopic car, and took a handful of average soil and shook all the hyphae out, then stretched them end to end, you could just ride the fungal superhighway for sixty miles," Frey said. These networks of fungal hyphae move water and food all through the soil, optimizing growth and survival in ways researchers are only beginning to puzzle out. The network grows among different plants, too, connecting individuals of the same, and even different, species. The staid, silent, orderly, discrete, noble bearing of the big oak that we see aboveground in the forest, standing seemingly unto itself, sedately alongside its neighbors, is nothing like the welter of interplay under way underground. Cooperation, competition, inhibition, mutualism, parasitism, it's all going on in the soil, along with a babble of communication by chemical compounds.

The biodiversity of life aboveground also has nothing on the suite of life in the soil. Where to begin? Consider the tardigrades— little water bears. Discovered fewer than 250 years ago, these tiny beings are related to insects but unique enough to be in their own category of animals. Colorless, white, brown, red, green, orange, yellow, and pink, they can be found in nearly every environment on the earth, including leaf litter and the top layers of soil. With their stumpy, clawed legs, and jabbing needlelike mouth part, they are ferocious predators of protozoa, nematodes, and even other tardigrades, and so tiny that as many as four hundred thousand of them can be found in a square yard of soil. They comprise a nation of more than nine hundred species and are capable of surviving temperatures from below zero to above boiling, intense radiation, and in suspended animation for decades without food or water. And what of the mites, the springtails, the ants, the legions of pale, blind dwellers with unpronounceable names of the damp, navigating only by scent and touch? Billions of bacteria are in just a single ounce of soil—it goes on and on. The environmental conditions also change in short distances. "You may have a little pocket with anaerobic conditions," Frey said. "A millimeter away, it may be aerobic. Trying to understand that dynamic is not impossible, but it's close."

So the big oak dominates its space, true. But it also supports a vast web of life and relies in turn on a menagerie of helpers, aboveground and below. With its crown in the wind and its roots in the teeming soil, the big oak connects earth and sky, and many millions of beings, and is home to each and to all. It is just one tree, and yet a whole world unto itself. How, I wondered, could I ever get to know it?

CHAPTER THREE

TO KNOW A TREE

I WAS TAKING the wool shirt off its hook in my office in the fall when I saw the inchworm, doing its inching-along thing, steadily up the wall. It had to have hitchhiked inside with me, no question. By then, this was nothing new. The longer I lived in the woods, the more I felt I was gradually not just observing the Forest, but becoming it. By Columbus Day, the Woods Crew had fired up the wood-fueled boilers that heated my house, and Shaler Hall. I watched the smoke puff and stream from the shed outside my office window, where the wood furnace toils, a perfect wind sock by which to gauge how much to layer up before heading out for a walk. The first thing I would hear every morning was usually someone in the Woods Crew heading up the forest road in the John Deere for another load of wood, starting the cycle of our days. The sweet scent of burning wood, cut from the forest, found me each morning as I walked out the door. The forest had gradually been coming inside, too. It started with that inchworm.

Next there was the fuzzy white caterpillar, cruising along the back of the couch. Then one day, working at my desk, a spider rappelled gracefully from my reading glasses. I was starting to get used to this. By late fall, the leaves started coming inside, too. The woods were at their loudest then, with deep drifts of crackling dry leaves everywhere. On the woodland trails as we walked, great mounds of leaves cascaded over the tops of our shoes like waves over the bow of a boat. Leaves blew into my house every time I opened the door. At first I swept them up. But like the insects, I eventually got used to them, too. Friends brought me mushrooms snapped from the woods on afternoon walks: apricot-scented chanterelles, giant chicken of the woods. I was thinking about the forest all day, photographing it, walking in it, reading about it, lying down for naps in it. The mushroomy, humusy scent of the forest was everywhere, and now it was seeping into the house along with the leaves. I was surrounded by the woods at all hours; every window in my life was filled with thousands of trees. But my imagination and gaze were fixated on just one: the big oak.

Beautiful to watch, late into November the oak's nut-brown leaves were still sailing off its branches and twirling to ground. I sketched the leaves as they cruised through the air, just to observe closely their spin, flight, and landing. They made up with flying finesse for their lack of color relative to their orange, red, and yellow neighbors. Oak leaves don't just drift like maple leaves in a lazy seesaw. Instead they twirl stem down to the ground. As the big oak slowly shed its canopy, I could see the tree better than I had in weeks. The rhythm of leaves falling all around as I gazed at the oak had a hypnotic effect, and there was something wonderful about staying still and quiet long enough to hear the impact of a single leaf hitting the ground. I noticed something else, too, by sitting still: the motion all around me. The blow of the wind was made visible by the swirling leaves it carried, as if on a stream of water, flowing, flowing around the trunks and branches of the forest, and pouring down from the sky. I would have to wait for flying snow to see again the atmospheric river of wind in which we are always so immersed—usually invisibly—so clearly.

But even amid all those trees, over thousands of acres, and living and breathing the woods, I still didn't feel as if I knew what a tree *does*. I observed the big oak daily and had by then watched it through four seasons. Yet somehow, it wasn't enough. What was it like up there in its branches? I decided that to know the big oak and its world well, I would have to change my vantage point. That's where Melissa LeVangie came in.

The tree warden for the town of Petersham, Massachusetts, where the Harvard Forest is located, LeVangie helps local residents make smarter decisions about their trees, and she is also the town's tree cop, taking out trees in the public right of way if they pose a safety risk. For all that, with a brass acorn necklace, a silver ring on one finger in the shape of an oak leaf, and her e-mail signature *sent from the trees*, there is no question where her sensibilities lie. A champion tree climber, she figured out in college that her passion wasn't forestry, but arboriculture: the profession of tending and caring for trees. She launched her career patrolling trees for invasive insects, working more than a hundred feet off the ground if need be, swinging from a battery of ropes, a harness, clips, and carabiners. I wrote to her a few months after moving in at the Forest, telling her that I wanted to get up in the big oak. Would she take me on a guided climb?

As the day for my first climb approached, I felt nervous, the thought of it crowding in from the wilds at the back of my mind where fear lies. But I was determined to try. LeVangie often spoke of her joy at being up in a tree—she climbs most every day for her job—and that made climbing the big oak seem at least survivable, even if not exactly fun. But as we got started, and I strapped into a climbing harness and clipped into a climbing line, I felt like a trussed turkey. I puzzled at my ropes and hitches and knots and thingamajigs, all of which I was quite sure were crucial, in a life-or-death kind of way. LeVangie had her blood type printed on the top of her climbing helmet, which I found both practical and a bit unsettling. We tested the harness, just to be sure, before leaving the ground, LeVangie holding on to the climbing rope as I leaned back in the harness, letting it receive all my weight. For better or for worse, all systems were go. This was the moment.

I looked up the rope and gave the friction knot a push. All of a sudden, my feet swung free, off the ground. As I felt myself hang in the air and saw the big oak's trunk crowd in close, it came back to me in a rush. Here it was, that feeling all over again, when I first climbed the cedar by our pond as a girl, grasping the fragrant, peeling cinnamon-barked branches to haul myself up. I was alone (that was important), and each time I paused and looked out and around, I still saw more tree above me and just kept on going. In no time, I was almost to the very top, the trunk narrowing until I could span it with my hands. The whole vista of my world of our house and pond suddenly resolved into a new perspective. I was at ease and thrilled at the same time.

This climb was the same thrill, but with a lot of help. The big oak has no branches at all for some forty feet; the only way up was on a rope with nothing but free fall all around. Unnerved as a novice never good at knots, I leaned hard on LeVangie's confidence and focused on the physical work of climbing, trying to just stay in my body rather than let my mind run off to worry. I huffed and heaved, no graceful first timer, as I went higher and higher. I looked over at LeVangie, effortlessly climbing alongside me, and finally understood a key thing I was doing wrong, in the way I was holding the friction knot on my climbing rope. I tried a new grip, pushed the rope away (as indeed she had patiently instructed so many times) to let the knot slide freely up with each push. I began to make more steady progress. Then, something magical happened.

With a twist and a heft, suddenly I was sitting in the big oak tree. On the first rise of branches, my relationship with this tree changed. I was in its realm. And how different everything looked from up here. The neighboring trees came so near with their branches, poking into any space the oak wasn't filling. The lichens were different up here—there were many more of them, a whole garden up in the sky I was never aware even existed. And who knew about that cavity, way up here in the trunk? Did anything live in there?

"Do you like birch snacks?" LeVangie said, handing me a black birch twig from her pocket. It tasted of wintergreen, fresh and cool. Chewing my twig, I looked around at the oak's branches, stretched out at my level in a great broad sweep. I felt a pure, unalloyed joy. I looked down to the ground—quickly—then away. "You'll gain that trust," LeVangie said from her perch on a limb across the tree's trunk. She encouraged me to relax, stretch out, enjoy myself. I leaned back in my harness, felt the nice stretch in my back. Sat back up and looked around. I saw the snow lying in patterns on the partly bare ground, rather like the lichen on the tree's bark. The shape of the stone wall resolved to a clear long line. Up here, the wind sounded different, fuller and more powerful. The swish of the white pine nearby was splendid, yes, like the sound of the wind at sea. For really, I was sailing the wind, from up in my crow's nest of oak.

We were up there quite a while, exploring. I finally stood up on the big bough I had reached and enjoyed a look around. How I loved the solidity of the tree under my feet, what a realm it commands, and with such authority. A connoisseur of trees, LeVangie said of the many species she climbs, oaks are the firmest. "One of the things I love most about oaks is how solid they are," LeVangie said. "It is like climbing an iron scaffold." Beeches are graceful like no other tree from up in their branches, she said, while other species have a nice give and bounce. She enjoys knowing each tree up close. "When you are in different trees, you get to watch every nuance and change, to watch the trees go to bed in the winter and come alive in the spring. The chickadees come and say hello," LeVangie said. "You hear different things up here, you can call the tree frogs, and they will call back to you." Now *that* I wanted to try.

One of the things I was surprised by is how many dead branches were up here, dried and crabbed, some studded with mushrooms, or even dangling, but just not fallen to ground yet. Especially distinct, too, was the oak's twist. From up here, that was as apparent as the pattern in a barber pole.

A lot of tree was still left to climb up above me, but I suddenly found myself exhausted, overwhelmed with the newness of it all, and ready to go down. Our descent was an easy controlled slide down the rope, with stops to break my momentum. The ground, once I reached it, felt soft and springy underfoot, after the sturdiness of that branch. My guided climb, I decided then and there, wasn't enough. I wanted to know this tree top to bottom, wanted to work in it, think in it, yes, be able to climb it myself. I resolved to climb it often, to feel and see its world change through the seasons. It would take a lot to know a tree, I was coming to accept, even just one oak.

Surrounded by trees in our lives, what do we really know

about what they are doing all day? They are our close neighbors, as LeVangie says, in nearly all of the spaces and places we live. Yet for all of humanity's long coexistence with trees, their mechanics and physiology are something we are still seeking to understand. The dawn of vascular land plants on the earth dates to more than four hundred million years ago, long before the first vertebrate animals. By ninety-five million years ago, many trees we would recognize today were flourishing.

From their earliest beginnings, all trees have retained certain characteristics. They are a woody perennial plant, with a single main stem, and branches a distance from the ground, crowned with a head of foliage. But there are always exceptions. Trees at high altitude, battered by severe winds and snows and their harsh environment, take knee-high forms shorter than many shrubs— but they are no less a tree. Trees have so many forms, life strategies, and textures. The bark of a beech is smooth as skin, while that of the oak is craggy and furrowed. The leaves of trees are a myriad of shapes, not only across species, but on the same tree depending on their position or life stage.

Trees have always fascinated people. Aristotle considered plants as rooted animals, at least metaphorically, and how right he was. They demonstrate the agency of the stationary. They are rooted in place and shaped by events around them: the growth of neighboring trees, prevailing winds, weather. But they are not passive. Trees manipulate their environment, exuding chemicals to deter pests and call in predators. They make soil, alter the hydrologic cycle, climate, atmosphere, and habitat. Trees move, breathe, operate a whole-body circulatory system, eat, have sex, communicate, expel waste, socialize, wage war, compete, cooperate, and create. Would that any of us could be as creative, productive, and responsive to our world, day in and day out, year unfailingly after year, with the quiet finesse of any tree.

Their form beautifully enables function. Branches hold the leaves aloft so they may soak up sunlight to make food. On the outermost edge of the trunk is the bark, protecting the tree. Just inside that is the inner bark, made of phloem cells: the first part of the tree's circulatory system. A thin layer of long pipelike cells, the phloem fit together vertically, end to end, the length of the tree's trunk and branches. Phloem carry the tree's nourishment from where it is made in the leaves to woody tissues and roots, where it is either used for growth or stored.

Next to the phloem comes the cambium, a thin, living layer of cells capable of dividing to form new phloem cells to its outside edge, and to its inside edge, xylem. This inner layer of xylem is the wood proper: the cells built with cellulose and lignin that support the tree. Xylem cells are also the second part of the tree's circulatory system: the pipes through which water moves up the tree from the ground, all the way to its crown.

At the innermost core of the tree is the heartwood: the dead, oldest heart of the tree, hardened and protected (somewhat) with chemicals the tree produces to thwart bugs and rot. A tree is many ages at once, from its ephemeral leaves and newly formed xylem and phloem cells and the perpetually living cambium, all the way to its long-dead core. The cambium is the living skin of the tree—this is why girdling a tree kills it without cutting it down. Girdling severs the living cells of the tree, cutting off both its nourishment and regenerative capacity.

The rings of the core David Orwig wrested from the big oak are arranged in light and dark bands. Both the light and the dark parts are xylem—woody tissue—but their color differentiation reveals another feature that is particularly pronounced in red oak. Its structure is porous—so porous you could blow air through a block of it. But as in everything else in the tree, that porous form allows a particular—and clever—function. Red oaks are not among the first trees to leaf out in the spring, which helps them to dodge a late frost. Yet they grow bigger and faster in a season than their quicker-out-of-the-gate neighbors. How do they do it?

That porous structure is the key. It is a superhighway for transporting water and minerals and stored sugars to the leaves in spring, to kick-start the tree's growth. This xylem tissue is grown every year in earliest spring, and in all trees is lighter in color and less dense than the band a tree makes the same year in summer, as it gradually ramps down growth with the declining angle, day length, and energy of the sun, until finally dropping its leaves and going dormant for the year in the fall. Each couplet of light and dark wood together reveals the tree's growth for one year. The trunk thickens progressively outward as the tree grows, with the oldest wood at its core, and newest just under the cambium. A fast-growing tree has big rings. A tree shaded out by its neighbors or stressed by drought, fire, or other insult has tiny, tight rings as it barely grows, perhaps for many years, until it is released by a change in its environmental conditions. This is why reading a tree core tells a story of the tree's life: how long it has lived, and when, over that time, it was growing fast and free, and when it was suppressed and hunkered down. The whys can be revealed from corroborating information, found in the cores of other trees in the stand, and the stories people keep and tell in newspapers, land use records, family correspondence, weather records, farmers' journals, and more. Trees are scribes, diarists, historians. They tell stories. They are among our oldest journalists.

In a tree's life, every year is different, but they all start the same. The tree's annual growth starts in the roots, when changes in temperature and day length in early spring trigger hormonal responses that cause root cells to secrete minerals and sugars into the xylem of the tree. Water follows from the soil by osmosis, which forces water up the trunk—sap rise. But that pressure isn't enough to move water all the way up the tree. Nor is the capillary action of the xylem. The two together make the whole system work, up through the miraculous mechanism of the leaf. Remember the pores on the bottoms of the leaves—the stomata. These open to make the tree's food—I'll return to that—and as they do so, the leaves lose water through the openings, by transpiration. The flow of water through the tree is a function, then, of the rate of this evaporation of water from the tree, through its leaves.

It all works because of the properties of water: a combination of attraction and stickiness. The hydrogen bonds of water are strong, holding the molecules together. The molecules are also sticky, with a positive and a negative end, allowing them to cling to one another. This combination allows the water molecules to move upward, in a continuous stream. The ascent even to great height is powered by the ever-smaller size of the transport vessels, from xylem in the trunk, to the tiniest portal of all, the stomata on the undersides of leaves, ten times finer than a human hair. This ever-tinier restriction of each subsequent transport vessel continually boosts the pressure inside the cells, carrying the water ever upward, and finally, up and out of the leaf as vapor, through those millions of microscopic mouths, into the atmosphere. The next water molecule in line moves up the chain, and the process continues, moving water up the tree even hundreds of feet, all the way from the soil. As much as 99 percent of the water taken in through the big oak's roots is lost through simple evaporation, or transpiration of water to the atmosphere from the leaves in just this way. On a hot, sunny day, a tree may transport hundreds of gallons of water from the soil to the atmosphere. Trees pump so much water into the air that they affect ambient humidity, cloud formation, the clarity of the air, color of the sky, and the weather.

The tree does all that pumping against the countervailing forces of gravity and friction, without making a sound or using a calorie of energy. The number of interconnected water transport conduits—xylem cells—can exceed hundreds of millions in the trunk of a large tree such as the big oak, and their total length can be greater than 200 kilometers, or some 124 miles. The speed of water flow up a ring-porous tree such as the oak is also impressive, on the order of twenty meters or sixty-six feet per hour. The forces involved are enormous; a tree will actually slim ever so slightly in shape on a hot, dry day, as the suction of evapotranspiration pulls in the sides of the tree like a drinking straw. The tree replumps at night, as it refills with water.

I once watched Julia Fisher, an undergraduate student on summer study at the Forest, as she worked on an experiment in the lab on xylem tissue of red oak to learn just how much pressure the cells can endure before they burst. On a laptop computer screen display I saw her insert a tiny needle into a single cell and blow air under pressure into it, to determine how much force it took to break the cell wall. Incredibly, the dial hit 350 pounds per square inch of pressure before tiny bubbles of air finally fizzed at the bottom of the red oak twig—more than four times the pressure of a bicycle tire.

But the elegance of its transport system is just the beginning of the wonder of what the big oak is up to all day, throughout the growing season. Perhaps its most impressive feat is photosynthesis. A complex process involving many reactions, photosynthesis takes place using the simplest of materials: just carbon dioxide, water, and the light energy of the sun. In the big oak, carbon dioxide—the tree's source of carbon, from which it makes sugar—enters its leaves through the stomata. The tree uses light energy from the sun to combine carbon dioxide from the atmosphere with water to create the simple sugar glucose, to be used or stored by the tree. Oxygen is a by-product of the process, released to the air by the leaves through the stomata.

The emergence of photosynthesis was a transformative event

on our planet. The heroes of the Great Oxygenation Event beginning some 2.3 to 2.5 billion years ago, humble oceanic cyanobacteria—still working away on the earth today—first created the oxygen that began accumulating in the earth's atmosphere, eventually reaching levels that could sustain plants, then animals, and even us. Oxygen levels have stabilized at 21 percent of our atmosphere today, thanks to the miracle of photosynthesis. Other gases continue to accumulate in our atmosphere, including carbon dioxide, creating problems I will return to.

The tiny transformer driving photosynthesis is the chloroplast, a specialized structure within living plant cells derived from ancient ancestors of cyanobacteria. It's that green pigment called chlorophyll that we see in chloroplasts, cyanobacteria, plankton, and algae that makes it all work. The naturalist Donald Culross Peattie writes, "Every day, every hour, of all the ages as each continent and equally important each ocean rolls into the sunlight, chlorophyll ceaselessly creates. Only when man has done as much may he call himself the equal of the weed. Plant life sustains the living world. More precisely, chlorophyll does so."

I first saw living chloroplasts packed with green chlorophyll for myself during a botany class I was taking at the Harvard University Herbaria in Cambridge. After a long lecture on leaf anatomy, we trooped to a lab in the basement of the herbaria, where we took out powerful microscopes and sliced bits of ordinary blades of grass into shreds to examine. I took this bit of green quotidian life, parked it on a glass slide with a drop of water, and topped it with a small plastic cover slip, using fine, flexible metal tweezers. I switched on the microscope's lamp, dialed in the optics—and stepped into another world. Little green dots were zipping around in the cells: chloroplasts. Who knew they moved? Here were the tiny things that run our world. And this was in just one shred of grass. The scale of activity that had to be busily under way in the thousands of leaves resplendent on the big oak dazzled me.

I thought the tree was just . . . standing there. That is true enough in winter, when the oak is dormant. But the rest of the year, its plumbing surges with food and water, coursing up, down, and out, in a living, interactive connection to the earth, sky, and sun. Its leaves are abuzz with light energy jolting chloroplasts into action. Its canopy is minutely nubilous with freshly made oxygen and drifting water vapor released from millions of pores in a mist so fine we cannot see or smell it.

At the center of the action is carbon: the building block of the tree's tissues. The net primary productivity scientists measure in trees is the result of a tree's activity through the growing season: how much carbon it sequesters in its tissues, rather than gives back to the atmosphere in respiration. The carbon is an essential component of the sugars the tree makes through photosynthesis, which the tree can allocate to growth or reproduction or tuck away for later. How apt that the word *wood* is derived from the Latin *lignum*—meaning "that which is collected"—for trees make themselves, year by year, by gathering sunlight, water, and air. The leaves the big oak pushes out in the spring are the result of the sugars it stored in its roots the previous growing season. In that way, the annual growth of a tree echoes its previous year's productivity. In a good year, a tree can put on a lot of growth and also pack away good energy reserves. The following spring, the oak can put on not only leaves and a big spurt of growth, but even reproduce. A mast year will only happen if the tree has a lot of energy reserves, from a fat summer the year before.

This has implications for people, particularly people who enjoy maple syrup. Researchers have learned that a mast year in the fall portends a crummy syrup season for maples in the coming spring. The trees will have raided their sugar stores to put on a big crop of seed the previous fall. That means the spring sap will be watery, and it will take more energy to boil it down for a decent batch of syrup. Once again, everything's connected in the forest, not only biologically and ecologically, but also temporally.

We don't see or hear any of this busy, ingenious activity, all this pumping and manufacture and storage and transport. We walk along below trees with no salute, not so much as a thank-you, as they quietly, ceaselessly go about their business of making life on earth possible. All through processes we not only didn't invent, but can't even imitate, that given the right environmental conditions are entirely self-sustaining and renewable in a single location, through hundreds and, in some tree species, even thousands of years. I can't think of any human or animal enterprise that is the equal of the day in, day out miracle of a tree. I am often struck when people ask me just what it is about this tree I was writing about that makes it so special. To me it is the miracle of the ordinary, as the big oak unspectacularly carries on its spectacular daily life. I can't believe what it *does*.

Once on a full-moon night in the fall, I set off up the forest road to the big oak, walking over the pleasingly sandy road in my slippers, bits of grit kicking into their open backs. Silver light spilled over every surface, and the crickets were singing. Mica flecks glinted in the lane, and there wasn't a bit of wind, or a cloud. The moonlight from the great dome of the sky cast long shadows from the big old sugar maples and red oaks that line the lane. The walk felt like a mission to a secret world, a night-light in my house casting a pretty, golden glow from one upstairs window. I hadn't realized until I took this midnight walk that you could see that bit of glow from the forest.

I made my way up and into the woods by moonlight, following the path I knew so well to the big oak. It was easy to spot even in the dark, the biggest tree in the stand it shares with yellow, black, and white birches, red and striped maple, white pine, and other oaks. The moon hadn't yet risen above the trees and its light beamed between the trunks. A few stars were visible through the leaves. They seemed to blink on and off as the trees stirred when a breeze came up, the leaves occluding, then revealing, the stars' shine. I could hear the hum of Warm Ants and the whine of the pump on a gas analyzer on the Barn Tower. But amid all that the Forest was still a forest, unto itself, supreme in its sovereignty. I thought of what it was that is so special about trees: their stately, parallel lives, so independent of our own—and potentially so much longer. Trees contain time.

More than a century stood here, in the big oak. Through two World Wars; farms gone to forest; a carbon and digital age begun and still surging—and the oak just grew steadily on. Not unchanged, but persisting to each next day. I thought of it shut down now for the night, a solar-powered miracle at rest during the reign of the moon. Could I sense the big oak and the lives all around, breathing, just like me? What if the respiration of trees was visible, diffusing into the night sky in a soft golden cloud? Perhaps then the interconnectedness of our lives, easy to forget, would be more visible, our shared, living breath and connection to this earth more apparent. I had never realized that trees have daily schedules as surely as we do, with activities particular to the hour, and resting cycles that correspond to day and night. Graphed on paper, a day in the life of a tree looks like a bell-shaped curve, from a quiescent dawn to its height of activity at noon, winding down to its repose at sunset. I lay under the tree to look up into its crown, the leaves shut tight for the night, like a house waiting for morning.

CHAPTER FOUR

A Forest, Lost and Found

THE SMALL WORLD I lived in at the Harvard Forest is a historical landscape, a replica, carefully curated and maintained, of the Sanderson farm, one of the core properties from which the Harvard Forest was created. To get just the right mid-eighteenth- to nineteenth-century look, Harvard Forest director David Foster in 1992 had a bit of the forest felled and a board fence installed around the resulting pasture. Then he had cows trailered in each summer to graze it, for it otherwise relentlessly grows back to trees. The result is a classic New England pastoral landscape: the white Sanderson farmhouse, carefully restored and used by visiting researchers and fellows, with the adjoining rolling verdant pasture and strolling fat cows. The board fence defined three sides of the pasture, and the view to the Forest on the fourth side was delineated from the pasture by an old, lichened stone wall.

The cows basked on the pasture's upper, sunny slopes, occasionally taking shelter in their wooden loafing shed in the middle of the field, with an apple tree just to the side. It's a gnarled old tree, with a bit of a lean. The cows were amazingly hardy though, staying out pretty much in all weather. In the early winter as the snows came on, they sat placidly side by side amid the white drifts, great, inert cowbergs, idling the afternoon away, sharing bodily warmth and watching the flakes fall. A stream runs through the pasture, feeding a small pond, its silvery rills on the downslope side flashing in the sun. Idyllic, bucolic, the pasture and woodlot beyond were a view beautiful in any season, at any hour of the day. In winter I spent every morning looking at this pasture from a glassed-in sunporch on the east side of the house, eating breakfast in my lap and having my morning coffee. In the warmer months, I was at a picnic table planted by the Woods Crew in the yard by the pasture fence, watching morning come, or evening, any day I was home, which was nearly every day. It was that beautiful. The sound of the cows chewing their hay was a meditation. I liked watching their Stations of the Bask, as the cows moved around the pasture with the sun. I could watch them for hours.

In this, I was far from alone, a fact worth wondering about. The cows, I noticed, had charisma. They were the first thing tour groups typically wanted to stop and look at when they came to visit the Forest, and they always drew smiles. People brought their kids by on weekends just to pet the cows through the fence. The pasture is small enough that it could just as easily have been mowed twice a year, but the cows were themselves historical reen-actors, co-opted into our living exhibit of a New England pastoral landscape. Using animals to defend the open meadow from the encroachment of the Forest was the whole historical point.

The Sanderson farmhouse had been remade many times over since its construction. Used by the Harvard Forest for its head-quarters when Harvard first acquired the research forest, with the construction of Shaler Hall in 1938, it was remodeled again into apartments for visiting researchers. The communal coin laundry was in the basement, and when I headed down there with my washing, I often prowled around looking for clues to where the footprint of the original farmhouse might have been. Perhaps the monster rock foundation for the main chimney that loomed in the gloaming was original, set in place sometime after Sanderson purchased the property in 1763. The house went through many iterations, eventually rebuilt to two stories with a mansard roof, where the Forest billeted its students. By the time I lived there, the mansard roof was long gone, and the farmhouse was named Community House. It was home during my stay to visiting researchers from Australia, Pakistan, China (twice), and roving academics from all over the United States. When I was looking into where to live while at the Forest, Community House quickly emerged as clear and away my first choice—one, for the interest and pleasure of being in the middle of all those people. But also for the historical resonance and beauty of the setting. And then there were the cows.

Since moving in at Community House I had found myself becoming quite invested in their comings and goings and daily routine. Their bellows, loud as foghorns, as they called for their morning hay, were among the first sounds I heard each morning. Usually the only person living on the property full-time, I eagerly volunteered to lob their hay bales over the fence on the weekends, when the Woods Crew was away. I advocated for expensive green bales, too, over the cheaper, brown rounds Wisnewski, of the Woods Crew, sometimes brought over on the John Deere, stuck on the tines of the forklift attachment like a giant shredded-wheat biscuit. The cows loathed this inferior fare and would not even follow the tractor when it arrived or take notice of the meal thus delivered. Instead they would stand and bawl at the fence, lobbying for better until Wisnewski roared away in the tractor and the cows figured they had no other choice.

Ironically, hamburger rolls—the white-bread mass-produced ones, with the texture of soft foam rubber—were the cows' favorite thing. Wisnewski filled me in on this, and the minute he did, I was off to the nearest gas station convenience store, buying the cheapest, gummiest, squishiest rolls I could find. I tossed them across the fence or hand-fed them, amazed at the cows' disturbingly large tongues, practically prehensile, long as a forearm, and rough as a cat's. My favorite was Norma, a bug-eyed Jersey cow with a weathered buckle-on collar like a dog's. She kept watch for anyone approaching the fence who might just be holding a treat.

Including Norma, the herd numbered six cows and two calves, both born since I arrived, one black-and-white, the other a glowing cinnamon brown. The black-and-white baby calf soon learned it could step right under the fence and help itself to the hay bales stockpiled for each day's meals. The other cows—including the cinnamon calf, who never quite figured this out—balefully looked on, wondering, I suppose, how much would be left by the time the black-and-white baby cow finished snacking. Its free-range walkabouts concerned visitors, prompting so many phone calls to the Forest receptionist that the Woods Crew finally put up signs on the fence declaring CALF WILL GO BACK ON ITS OWN.

This whole business with the cows caused me to think a lot about what it is that made them such a draw, not only for me, but for so many visitors. Is it that we are so starved for any interaction with animals, beyond a cat or a dog? But I think it is something else, too, an unarticulated longing for a perceived-lost golden pastoral age of simplicity, verdure, and bounty. Amid the seductions of the beauty of the pasture, the white farmhouse, and classic stone wall and woods, it was easy, at least for me, to be quite wrong about what life had actually been like for the people who lived here when cows mattered. Not for historical reference and maintaining the scenery, but as part of precarious family livelihoods.

Wisnewski was always quick to remind me of this when I waxed poetic about the cows. Big and barrel-chested, Wisnewski stood tall as he talked, looking down from his silver-wire spectacles to meet your eyes directly, and holding steady, penetrating eye contact as he spoke. I rarely saw him without a hat, usually a much-faded John Deere cap. He didn't mind skewering theories and plans at the Forest for a revival of small-farm agriculture in New England with a reality check earned by hard personal experience. "Everyone wants to plant a garden, but no one wants to weed it" was one of his dicta.

"When we were growing up, no one admitted they farmed," Wisnewski told me one perfect June morning, the view to the pasture a lush symphony of green grass, golden sun, and blue sky. But that was just a visitor's view of a re-created landscape—not the reality of the farm life he had known. In his day, farm kids were looked down on and teased at school—if they were in school at all, Wisnewski said. "You didn't hear the sound of a bat at a ball field in summer; kids weren't playing, they were working." His eyes narrowed as he recalled one kid hung by his shoulders in a well by his father, until he came around to doing his share of the chores. Life was controlled by the bank holding the debt on the farm. And if the bank came calling, the desperation could result in that outcome too often reported in the obituaries of small-town papers: "Died at home." Code for suicide.

Yet one bitter day that winter, with colder temperatures overnight in the forecast, I heard Wisnewski say something I had a hunch about already. He and the Woods Crew had been out in the pasture for hours, trying to lure two last cows into a trailer to get them back to Wisnewski's home barn for the rest of the winter, before the coldest weather came. The cows did not want to go; it was an all-hands roundup. I broke out the hamburger rolls while Wisnewski herded the cows, and his colleagues on the Woods Crew helped steer them to openings in the fence. The cows were wily and spry; they would seem to cooperate, then at the last minute, as they approached the trailer, shy away and zoom just out of reach across the pasture once more. This went on for a while. Wisnewski suggested taking a break to let everyone regroup and headed into Shaler Hall for coffee. The sun was bright on the snow, the cows entertainingly good at their evasion, and this fine winter afternoon's outdoor test of wits more fun, hands down, than anything going on at the computer screens aglow in the many offices at Shaler Hall. "You know the truth is, we love it," Wisnewski said. They were back at it soon after, finally pulling the stragglers into the trailer with ropes. They teased me as they worked that it was all my fault the cows didn't want to go, after all those hamburger rolls and generally making their reenactor farm deal a little too sweet to leave. When I headed back to the house at dusk that night, the pasture seemed empty with the cows gone.

Surviving here had never been easy. When the newcomers arrived in north-central Massachusetts in about 1700, they found the land they called the New World was not new at all, but the home of Indians who had lived on these lands for thousands of years. The local native people subsisted not as agriculturalists, but by fishing, hunting, and gathering. They had no fixed villages; wigwams of saplings and bark enabled ready movement from one place to another. Looking at the history told in ancient pollen grains and sediment layers found in mud cores taken from local ponds, there is no sign of land clearing or fire. "This remained a forested landscape, dominated by natural forces," said David Foster, the director of the Forest, and one of its leading historical interpreters. "That is not to say they had no impact, but they were hunting, collecting material, their big resource was nuts, acorns, hickories, and chestnuts. They were very flexible, adaptable, and they interacted with other groups. Including Europeans." All known early-explorer descriptions of native villages document European goods. From the 1400s to the early 1500s, they find Indians with metal goods. The records of Spanish whaling vessels document Indians wearing pantaloons, and the records of early fishing and whaling by Basque, Portuguese, and Spanish people along the coast report the same, Foster said. Maize arrived late to New England, and the Indians cultivated it marginally, Foster said. But given the growing European presence, "It was inevitable that New England would become agricultural."

In the early days of Petersham, the first new settlers lived subsistence lifestyles, clearing small farms between about 1730 and 1760. It was crushing work. Those lovely stone walls I so enjoyed are a heroic testament to brute human and animal toil. Fencing the seventy-five acres of the original Sanderson farm entailed moving an astonishing ten million or so pounds of rock. Some of the walls were thick, to keep the cattle in the pastures. Other double walls defined plowed fields, with the new rocks heaved from bare, frostbitten soils each spring tossed between the two walls in a makeshift rock dump. The town of Petersham has

436 miles of stone wall—about 12 miles of stone wall per square mile. The bigger stones too heavy to lift were pushed to the base of the wall by yoked oxen, and the rest piled on top by hand. The total amount of stone used in all of the walls equals the volume of the largest pyramid at Giza. The sheer toil for their construction— never mind all the other work to be done as well—staggers modern sensibilities. Typically every farm would include about three miles of stone wall, with the work of building all that interspersed with clearing land, building, planting, harvesting, mowing, raising livestock, selling produce, cutting and hauling firewood, and breaking out roads in the deep winter snows. Women, in addition to family tasks, would have picked their own wool, carded and spun it, made cloth, cut, made, and mended clothes, made soap, bottomed chairs, braided rugs and woven carpets, made quilts and coverlets, plucked geese, milked cows, and gone visiting on foot—if they ever had the time. But spare time usually was spent earning a little cash making shoes, or weaving palm hats for sale by local merchants that provided the fronds imported from Cuba. This was primarily women's work, part of a robust "putting out" economy that deployed any spare time to earn scarce cash. The women of Petersham wove a spectacular 130,525 palm-leaf hats in 1836 alone. A typical winter's work was 250 hats, woven for ten cents each.

In Petersham the first divisions of common land titled to private tracts were given as payment to returning fighters from the French and Indian wars in lieu of cash. The General Court of Massachusetts transferred the first seventy-two Petersham land grants totaling twelve hundred acres in 1733, and the town was incorporated in 1754. Jonathan Sanderson purchased a fifty-acre parcel in 1763, beginning a family farming tradition that would last into three generations and make theirs one of the most prosperous families in the county, with the farm continually expanded to a total of 329 acres when it was finally sold in 1845.

Sanderson carved the farm from hills and valleys covered with a primeval forest of red oak, red maple, white ash, paper birch, and scattered white pine and hemlock. There was more beech and hemlock than in the woods today—shade-tolerant trees, thriving in the heavy canopy. American chestnut, important both to wildlife and people, also once towered in the New England woods. Sanderson had to drop trees with an ax, limb and remove brush to leave space for his crops, and plant between the stumps and rows of fallen trees. But as Foster and O'Keefe explain in their book, *New England Forests Through Time*, the primeval forest that settlers encountered in the 1700s was not only an empire of giants. Natural disturbances, from hurricanes to summer microbursts, winter ice storms, pests, and pathogens, left their mark in forests comprising an always-changing mosaic of trees, from young whips to venerable old growth. Part of what is missing in the Forest today is this range of diversity in age class—the truly large, old trees are long gone. Missing, too, is the complex, dense understory of downed trees in varying states of decomposition. Journals of travelers in the Northeast in the eighteenth century provide a glimpse of what these forests looked like before European settlement.

Writing in 1743, John Bartram describes the forests he encountered in his journey through Pennsylvania en route to Lake Ontario, Canada:

> We rode through a grove of white pine, very lofty and so close that the sun could hardly shine through. We set out on a northeast course and passed by very thick and tall timber of beech, chestnut, linden, ash, great magnolia, sugar birch, sugar maple, poplar, spruce, and some white pine . . . the very uneven, overfallen trees, turned up by the roots of prostrate timber, hence it is that the surface is principally composed of rotten trees, roots and moss, perpetually shaded and for the most part wet. What falls is constantly rotten, and rendering the Earth loose and springy. This tempts yellow wasps to breed in it which were very troublesome to us throughout our journey . . . We continued through a great white pine spruce swamp full of roots and an abundance of old trees lying on the ground or leaning against like ones.

He also noted gooseberries in abundance, and woods "crowded with wild pigeon"—the now-extinct passenger pigeon.

Confronted with such an outsize nature, settlers seeking to clear land and establish farms fought back with an old and reliable ally: fire. Burning of forests to clear ground for agriculture was so pervasive that on May 19, 1780, settlers panicked at what came to be known as the Great Dark Day. An atmospheric inversion combined with so many acres of burning woods meant the sun was blotted out by the smoke, and daylight so darkened that candles were needed to see as a preternatural gloom settled over the New England landscape. Many assumed that Judgment Day had come. One farmer wrote:

> To view nature dressed in her mourning attire: The earth enveloped in darkness: The husbandmen returning from their fields in great surprise: The midnight sentinels crowing in answer to each other: The dismal din of peeping frogs: The night-birds singing forth their dreary notes: The beasts grazing in wild consternation: Every countenance seemed to gather blackness: Yea, a dismal gloom which filled the beholder with fear and astonishment.

This extensive burning was just part of the land clearing process. Massive stumps left behind had still to be clawed from the earth, before it could be forced to accept the plow. In New England, oxen dragged stumps to the edges of fields, where they formed formidable fences that endured for years. The great naturalist and writer Henry David Thoreau (1817–62) delighted in the stump fences his neighbors built. He wrote in his journal on November 11, 1850:

> I am attracted by a fence made of white pine roots. It is almost as indestructible as a [stone] wall and certainly requires fewer repairs. It is light, white, and dry withal, and its fantastic

forms are agreeable to my eye. One would not have believed that any trees had such snarled and gnarled roots. In some instances you have a coarse network of roots as they interlaced on the surface perhaps of a swamp, which, set on its edge, really looks like a fence, with its paling crossings at various angles, and root repeatedly growing into root . . . so as to leave open spaces, square and diamond-shaped and triangular, quite like a length of fence. It is remarkable how white and clean these roots are, and that no lichens, or very few, grow on them; so free from decay are they.

Charmed as he was by the stump fences, the continual cutting of trees all around him vexed Thoreau. "This winter they are cutting down our woods more seriously than ever . . . Thank God, they cannot cut down the clouds," he wrote in January 1852. While we think of him as Pan with a pencil, living in a sylvan arcadia, Thoreau wrote at the height of deforestation in New England. Settlers had been cutting down the forest for a century by the time he settled in for his year in the cabin at Walden. So wholesale was the destruction by then that George Perkins Marsh (1801–82), the great conservationist, warned in his landmark book, *Man and Nature*, "The earth is fast becoming an unfit home for its noblest inhabitant, and another era of equal human crime and human improvidence and of like duration . . . would reduce it to such a condition of impoverished productiveness, or shattered surface, or climatic excess, as to threaten the depravation, barbarism, and perhaps even extinction of the species." That was in 1864.

In his book *Common Landscape of America*, John Stilgoe writes of the horror with which European travelers also observed the burned and maimed woods of "husbandmen" struggling to render farms from forests: "Travelers watched Americans burning great heaps of felled trees or harrowing seeds thrown among stumps and scorned the stupidity of American husbandry," Stilgoe writes. William Strickland, an English agricultural writer traveling in western New York, was equally appalled, writing:

> The scene is truly savage. Immense trees stripped of their foliage and half consumed by fire extend their sprawling limbs, the parts of which untouched by the fire, now bleached by the weather, form a stronger contrast with the charring of the remainder; the ground is strewn with immense stones, many of them of a size far too large to be movable, interspersed with the stumps of the lesser trees which have been cut off about a yard from the ground.

Tall stumps were not, Stilgoe writes, a matter of laziness or carelessness, but part of a carefully choreographed, multiyear strategic stump-removal plan in which after a year or two men chained an upright lever to the trunk of the stump, yoked oxen to the upper end of the lever, and wrenched the stumps out of the ground.

Or, rather than the harder work of chopping them down, settlers killed trees by girdling, cutting a trough into the bark all around to sever the living tissues. Trees were left to die in place.

While less laborious, the practice had its drawbacks for the farmers working underneath. Branches dropped to the ground often without warning, as the trees gradually disintegrated.

But what the horrified critics looking onto this charred, hacked, and dying landscape did not appreciate was that the settlers were making use of snaggletoothed, stump- and rock-studded land as best they could manage, in their struggle to feed and care for family and livestock even as they were wresting a farm from the forest. They lived off the land from the start.

For meals, the Sanderson family would at first probably have made do with berries, chestnuts, wild turkey, passenger pigeon, Indian corn, beans, potatoes, pumpkins, and bread of rye flour mixed with Indian-corn meal. The family's standard drink would have been hard cider, rum, or whiskey. In his book *Drink: A Social History of America*, Andrew Barr reports that settlers drank little if any water or milk. Water was often not pure and was regarded as a drink only for those who could afford or obtain nothing else. Milk spoiled. Moreover, people believed that drinking spirits provided strength and replenished warmth in the body's core lost through sweating.

Work followed the seasonal year: spring plowing, fencing, building, and planting; midsummer plowing and hoeing potatoes and planting corn, mowing and raking hay, picking fruits and berries. Autumn was for preparing for winter, threshing grain, getting apples and making cider, digging potatoes, and harvesting corn. After the first frost a fat ox or cow may have been slaughtered, and in winter, perhaps a steer. With the coming of snow it was time to break out roads and sled firewood. The coldest months were for work in the barn, doing indoor repairs. Heat was entirely from wood, tremendous quantities of it.

William Cronon in his book *Changes in the Land* notes that the typical New England household could have consumed as much as thirty to forty cords of firewood a year, or a stack four feet wide, four feet high, and three hundred feet long—requiring the cutting and burning of more than an acre of forest every single year. Trees and wood were at the center of every settler's world, and every farm had its woodlot, as essential to the household as cleared land. Biscuit and bread bakers also had to heat their ovens, and the numerous pig-iron foundries and forges needed great quantities of fuel, largely charcoal. Potash, used for soap and gunpowder, was made from the ashes of burned wood. And wood furnished the frames, planks, and boards for farm buildings, furniture, and dwellings, fencing, and tools, from oxcarts to plows, harrows, sleds, hay rakes and forks, flails, and shovels. Even table dishes, bowls and spoons, milk pans, and buckets were made of wood.

Sanderson and his wife, Molly, tilled twenty-six acres of cropland and cleared forty-nine acres for pasture, according to an 1806 map of their farm, by then seventy-five acres. A farm that size would have supported a pair of oxen, a dozen sheep, and a half dozen cows, plus a few pigs and chickens. Their manure would have been the only fertilizer for cultivated crops: corn, wheat, barley, rye, oats, buckwheat, and garden vegetables. Between 1791 and 1830, the cleared acreage in Petersham greatly increased, and expansion continued to midcentury as the local economy grew. By then, Sanderson's son John was not only a farmer, but a cattle drover, real estate speculator, and tanner. He prospered until his accidental death in his barn, July 25, 1831, while taking a pair of unruly oxen off the cart tongue. Lydia, his wife, took over the farm until their son, John Jr., then just sixteen, was ready to step up to the task.

The Sanderson tannery was on a stream just over the hill from where the cows now stroll the re-created pasture. If you know where to find it, the tannery site is still there to explore. Its location in the woods is not obvious, and Foster had to guide me through the site one late-autumn day before the snow flew for me to appreciate just what I was seeing in the old stone foundations poking up from the leaves. Not yet living at the Forest, I had driven out from Cambridge to talk with the Forest's director for the first time and arrived, laptop open, for what I thought would be a standard sit-down interview. Which proved only that I did not yet know Foster one bit. As we settled into his office, I started asking questions and avidly typing away, but in just a few minutes he asked, "Aren't we going to go for a walk?"

Director of the Forest since 1990, and on the faculty at Harvard since 1983, in his fleece vests and down jackets Foster looked more the hiker than the Harvard don. Any day in any weather, I'd see him swing his snowshoes over his shoulder, or a camera on its strap, to go out into the woods for a leg-stretcher and look-around. His colleagues were used to Foster's appearing on foot for site visits deep in the woods we arrived at by truck, and watching him melt into forest paths afterward to make his way back to Shaler Hall, content in his own company and the trees. He is a botanist and ecologist by training, and while some scientists are reluctant to bring their expertise and research into the public and political sphere, he has intentionally pushed it there. Foster had created a team on staff charged with bringing science to bear in public policy discussions, from regional land use and conservation to federal regulations on coal-fired power plants. He hadn't been afraid of scrapping close to home either, urging Harvard to take the golf course next door to court over illegal tree cutting on the Forest's property, then acquiring the property as part of the settlement. The day of our walk we pushed through barbed-wire fencing to see that property, already grown over after just one season to scruffy meadow grass. Foster and the Woods Crew's John Wisnewski were turning the place back to cow pasture, a pilot project in the Forest's evaluation of the potential for small-scale, local agriculture to conserve land from development. The experiment was also part of an expanded research vision for what Foster had already started fondly calling the Harvard Forest and Farm. As we turned around from the meadows and headed back to the woods, I asked Foster if he golfed. "Never in my life," he said with a level stare.

We pushed into the trees, looking for the old tannery site. "Here you can see where they flooded the marsh to make a pond. The dam was there," Foster said. "You can see the changing relationship of people to the land. He was a farmer," Foster said of John Sanderson, "but he also owned a mill. People thought it was just a gristmill, no one thinks of a tannery. But it was." He pointed to the millstone, used to crush hemlock bark, then loaded in vats to release its tannins, to condition some one thousand cowhides a year.

Petersham was thriving by 1840, reaching its all-time high in population of 1,775 residents, with about 60 percent of the land cleared, mostly for pasture, meadow, and mowing, and about 4 percent tilled for crops. An expanding road network grew to an intricate system of connecting and feeder roads. In Petersham, hardly an acre of land was more than a quarter mile from some kind of road. The establishment of this growing road network, and regional toll turnpikes, helped open up markets to Boston and towns in the Connecticut Valley. By 1850, an inventory of businesses in Petersham included two tanners, two gristmills, ten water-powered sawmills, one steam-powered sawmill, a hat-pressing factory, one bleachery for whitening palm leaves, one stream mill to press hats, a ladder factory, a tavern, two stores, one grocery, eight blacksmiths, four boot and shoe makers, two physicians, two butchers, two stonecutters, one carriage maker, two wheelwrights, one cooper, three painters, one millwright, one tailor, and one tailoress.

The Sandersons and others like them had put their money back into their farms, buying more land and improving their homes. The Sanderson farm remained intact until Lydia sold it in 1845. Clearing of the forest primarily for hay and pasture peaked at an amazing 70 percent of the land in Petersham by 1885. But changes beyond those opened horizons would alter New England agriculture forever: construction of the Erie Canal.

Opened in 1825, the canal gave farmers utilizing superior farmland in what is now the Midwest access to the populous and growing markets of the East. Massachusetts' population was surging, but a redistribution of people was under way. Farmers started moving away to better prospects, including wage work in the growing towns and cities, as the mill economy grew. Hill towns lost out not only to emigration, but to valley towns with better soil, industries, and railroad service, notes Howard Russell in his history of three hundred years of New England farming, *A Long, Deep Furrow*. Petersham's story was like that of so many other New England towns: At least half, and maybe more, of the open land in Petersham went out of farming. Partly, this was due to farmers' leaving agriculture altogether. Others did the math on pastures exhausted by grazing and let the land grow to white pine for harvest—while feeding their animals cheap feed grain from the booming Midwest farms. In Petersham, agricultural production plateaued between 1875 and 1905. On parcel after parcel, the open lands of Petersham, so hard-won, were allowed to grow back to trees—including the big oak by that stone wall.

One frozen winter morning, up early and waiting for daylight, I found myself thinking of Sarah Luce Mann. Born in 1801, she lived on Doe Valley Road in Petersham, not a mile from where I now sat. The snow was blue-white, and the sky a long, clear view into infinity. As the first pale light came, the black walnut tree outside my kitchen window stood in spidery silhouette, its symmetry a fugue of refrain and repeat. To the west, the board fence stitched the white snow on the pasture. I looked to the upper pasture, where the sun always comes first, but, no, not yet. I had been reading Mann's diary at the suggestion of Elaine Doughty, curator of the Harvard Forest Archives. In those quiet hours before daybreak, Mann's words were in my head. Did she relish this same secret, quiet waiting-for-the-light-time, too, looking at these same views of snowy pastures and woods? Or was she already at work?

Probably: in her diary, kept from 1861 to 1862 and written in her neat, careful hand at the end of each day, every minute of her life is a verb. Washed, scrubbed, churned, knitted, mopped, ironed, sewed, carried. And her all-purpose verb, *chored*, as in "I chored all morning." Making soap, making candles, upholstering chairs, mending. She uses the word *weary* a lot in her daily accounts. It is often underlined, sometimes more than once. She was the mother of six children, only two of whom survived to late adulthood; two died in childhood. One son died of something she called "brain fever"; a daughter was dead by age twenty-two, no cause given; another son lived less than two years; and another went off to fight in the Civil War in 1865, but died of fever before seeing action.

I spent a snowy afternoon at the cemetery where three of her children are buried. "Short pain, short grief, dear babe, was thine," reads the inscription on the somber black-slate grave, grown with lichen, for Asaph, died January 25, 1832, at age just one year and three months. "Alas how changed that lovely flower, Which bloomed and cheered my heart; Fair fleeting comfort of an hour; How soon we're called to part," read the gravestone for her son Richard Baxter, dead by age ten.

I was reading her diary, and visiting the cemetery, to better understand what life was like for people who lived here when these woods were pastures and farms. I wanted to pull back the wool of abstraction and romance from the history of this landscape. Her diary put new focus, too, on the reality of the intimate connection to nature that was her daily life. Life amid nature and on the farm was not for her the walk in the woods I enjoy or fun with the cows. It was a horrendous, unrelenting amount of work, wringing a living from the land in the era of horsepower and firewood.

Take Mann's week of April 16, 1861. Fort Sumter in Charleston, South Carolina, had been attacked by Confederates on the twelfth, and the Civil War had begun. President Lincoln issued a proclamation calling for seventy-five thousand militiamen, and Virginia, Arkansas, Tennessee, and North Carolina were about to secede from the Union. The agricultural economy in Petersham was unraveling. Great technological change was remaking markets, with the Erie Canal and railroad transport compressing the realities of time and space. But for subsistence farms such as Mann's, all this modernization and change didn't make life any easier—just more uncertain.

"The war news is much talked about. When will the Prince of Peace take the Kingdom," wrote Mann on April 16. She continued, "Bringing home casks of potash. Going to town for a load of brick. Made brown bread. Wheeler has moved a great rock where he is going to make a wall. Lumping butter to prepare for market. Mending stockings." Part of the stress in her household, and in many other families like hers, was the need to share tight quarters with several generations. She writes often of her eagerness for her son Henry to get on with his work of building a home of his own: "I shall be very thankful if I can have a room and quiet resting place, I hope for a little quiet, we have lived in confusion for some time." On May 1, she writes, "We have been getting out lye, preparing to make soap. Another day of trial is almost gone." On May 2: "I have been making soap, Harriet put the clothes out to dry, making brown bread to bake this evening. It is now past 8, and I am very, very weary."

On May 3: "Have not felt well, done but little." May 4: "Finishing up the soap. I am pretty thoroughly tired but another week is in the past and I am glad that things are no worse. I still hope for the land of rest, if I may be fit to enter there." On May 5: "Two men visited and talked of war, surely we live in perilous times. O happy day when wars shall cease and Earth will be filled with peace." May 6: "Made pair of pillow cases. Henry is getting along slowly on his building, I hope he will get into it before long, for I want my room very much indeed. Sewing shirts, and lumping butter." May 9: "Drawing out manure with steers. Planted cucumbers. Drawing out rocks. I have done a few chores, sewed, made some cake, fried some pancakes. A little on the whole accomplished but the day is in the past, one less for a weary pilgrim to wander." May 10: "Have done a large white washing. Taking steers to pasture. I am weary but I hope to rest [underlined twice]. Carried a dozen pounds of butter to Athol [neighboring town, five miles distant]. Getting wood. I have been choring and ironing. This week is past."

May 14: "Cloudy, damp, drawing manure with the old horse. Planting potatoes. Chimney for Henry finished." May 15: "Baked apple, custard pies. I am weary this evening." May 21: "We had some dandelions for dinner which took some time to prepare. Sewed some overalls. Done chores. I feel weary." May 23: "Corn planted, fixed fence." May 27: "We have worked hard and the day is gone." May 29: "Time is on the wing around me, whether I do little or much." June 5: "Last night Henry lost a calf jumping down a high place, surely they are unlucky about calves. Weeding corn. Rinsed clothes in the spring, spread to dry in the usual place. Another weary day is gone. Making cheese. Traveling by horse and buggy to Athol. Scything and mowing, raking hay." July 21: "First string beans for supper." August 9: "Fixed old arm chair with stuffing and cover. Knit three pairs of stockings. I have done what I could and another week is gone."

On it goes like that, week after week, the choring and the weariness, and worry of the war to which she would lose another son. Mann was sixty years old when she wrote this diary. She kept a diary seemingly only that one year.

Sarah Mann lived until November 14, 1883. Her house on Doe Valley Road, though much remodeled, still stands, and I went to see it one blooming day in May. I wanted to get a sense of the views from her windows, to see the hearth she had swept, the wide board floors she had walked. And there they were—the hearth, the floorboards of heart pine, more than a foot wide. The house stood on a foundation of solid granite blocks. The owner, Donald Flye, eighty, just back from his bicycle ride, greeted me warmly. He had found Mann's diary up in the attic and gifted a copy of it to the Harvard Forest, where his wife at the time worked. A note to the file in the Forest archives enabled me to find Flye and the house, built in the 1760s. We sat together at a plank table in the front room, a woodstove now where the open hearth had been. We looked through stacks of old papers, including family trees handwritten in pencil on yellowing paper, and old newspaper clippings: Flye was a local-history buff.

As I read and sifted, I realized that William Mann, a nephew of Sarah's husband, had lived in this house from the age of six to adulthood. After his marriage and his move away to New Salem and Barre, William returned to Petersham and carried on the John Sanderson farm for a few years, before finally purchasing it in 1845. With a jolt of excitement I realized Sarah Mann, whose life I had become so caught up with, had surely often been in my house, too, visiting William at the Sanderson farm. He was just one of a succession of owners of the farm, until it was acquired by Harvard University.

By the time the Harvard Forest was founded in 1907, many abandoned pastures and fields throughout the area were seeding in almost entirely to volunteer white pine. The pines, mature for cutting, were growing on land that just sixty years before had supported prosperous farms such as the Sanderson place. Large quantities of marketable white pine became available between 1900 and 1920, and another economic era, this time of logging and milling, began. Clearing these trees—the second-growth forest—opened the next chapter in the life of these woods, parts of which next regrew not to white pine, but mixed hardwoods. Including the big oak.

That oak has matured during the third transformation of this landscape. It grows not in the primeval forest felled by Jonathan Sanderson, nor the second growth of white pine, sprung up after Sarah Mann. It would be the third growth of mixed hardwoods that the summer people found so attractive, as the wealthy from Boston and surrounding communities began using Petersham for lovely homes, tucked along the town common, and interspersed among the few remaining farms. The mixed-hardwood forest regrew all around them, interlaced with the stone walls of a former era. This is the Petersham I had come to know, a place of gracious old homes, open pasture vistas along the main road to town, a green town common with its gazebo, white-steepled church, and lively country store. There is endless walking in every direction in a classic New England landscape of forests, meadows, ponds, swamps, and brooks, much of it in permanent conservation. Incredibly, the landscape that was not that long ago largely cleared is today about 90 percent forested. As I walked these woods, often my imagination lingered. I thought of the people who had dug the old cellar holes I saw amid the trees or, as I passed through a break in a tumbled stone wall, who had once installed a gate. I saw traceries of the foundations of barns and mills and homes, and the ruins of dams on streams now running free. Witnesses to the history of this landscape, just like the big oak.

As it sank its taproot by that stone wall, the number of commercial establishments in Petersham had by 1905 dropped to four, and its total power equipment was reduced to two steam engines and three waterwheels. The last local palm-leaf-hat factory, in nearby Dana, burned. By the time the oak was a seedling in 1910, the population of Petersham had plummeted to 757—not many more than the 707 felling the primeval forest when the first census was counted in 1765. Meanwhile the oak gathered size, pushing up to the sun. The third forest was just starting to flourish. But then, these woods would be transformed yet again.

CHAPTER FIVE

TALKATIVE TREES

Up in the dark, I made breakfast for a day working in the woods: eggs scrambled in butter in the cast iron fry pan, a big hunk of bread split and toasted, and the whole mess layered up with mayonnaise and *chiles en adobo*. I stuffed my pack with a thermos of hot coffee, snacks, and a monster peanut butter sandwich and was out the door, over to Shaler Hall. Lights in the windows glowed yellow in the still-wan morning hour. But the morning sky was clear and bright, and it was a delicious September morning, cold enough for a jacket. Perfect for a day of fieldwork in the woods.

I headed over for coffee in the common room at Shaler Hall, with its oil painting of the selfsame Nathaniel Shaler, one of the Forest's founding fathers, presiding from over the mantel. Deep leather couches and racks of journals and natural history magazines beckoned, but they were not in the plan for today. With Audrey Barker Plotkin and Ahmed Saddig, an ecologist from Sudan visiting Massachusetts to earn his Ph.D., we were a small but hardy field crew. Gathering our gear, we loaded the car, stuffing the back with oversize measuring tapes, data sheets in a metal carrying case, and what would prove to be one of our most important tools: big stubs of blue chalk. Saddig settled into the backseat with his statistics textbook, and Barker Plotkin slid behind the wheel.

I felt a sense of history. For how many years had Harvard Forest researchers gathered like this in the early hours to get a start on a day in the field, pursuing what the woods had to teach? Especially these woods: we were headed to Pisgah. At the Harvard Forest, this one word—short for the Harvard tract in the Pisgah State Park, in southwestern New Hampshire—evokes a research obsession under way for decades. For the record of what happened to the woods after New England's signature hurricane was recorded there, as nowhere else.

New England is no stranger to storms, which occur frequently and often violently. They are a defining force in forest dynamics and the life of the people. The earliest written records of New England relate storms "very strang and fearfull to behowld," as Governor Bradford of the Plymouth Colony wrote in his journal on August 15, 1635. But even so, the 1938 hurricane was historic in its violence. It blew through the heart of New England with sustained winds of 121 miles per hour and gusts of 186 miles per hour logged at the Blue Hill Observatory near Boston—the strongest winds ever recorded in the region. Before it was over, the hurricane claimed 564 lives. Damage to fishing fleets was catastrophic and flooding devastating. Providence, Rhode Island, was awash under a storm tide of nearly twenty feet.

The storm was born in mid-September, in the tropical Atlantic, near the Cape Verde Islands of Africa. As the storm encountered a stream of tropical air north and east of Florida, it increased its cyclonic action, growing to a three-hundred-mile-wide sucking whirl. By the morning of the nineteenth, the Boston paper carried the news of a hurricane disturbance in formation off the east coast of Florida, traveling westward at the rate of forty miles an hour. About three P.M. the next day it hit Long Island in full force, then pounded the coast of Connecticut and Rhode Island with forty-foot waves. But this storm was still not finished; it smashed its way through New England, heading north at about fifty miles per hour. Mabel Cook Coolidge in her book *History of Petersham* reports that at three P.M. the skies were clear after days of heavy rain that had saturated the ground. The citizens of Petersham were walking the streets, enjoying a perfect sunny day, while the oncoming hurricane was pummeling the south shore of New England. Then it hit. "For two hours it raged," Coolidge recounts:

Windows crashed, doors blew in or went sailing off, roofs trembled, wavered and came off, shutters and piazza chairs were blown so far away they were never found, chimneys toppled on crashing roofs. Trees bent lower and lower, then in despair turned up their heels, taking with them the wires of the telephone and electric lights. Chicken houses were turned end over end, several times shingles and slate blew from roofs like flocks of pigeons, all this accompanied with the worst downpour of rain, mixed with sand, and leaves . . . The buildings, especially those painted white, looked as though they had been scorched by fire, because of the mash, composed of pulverized leaves, sap and sand that was beaten into the paint by the wind.

Rail service between New York and Boston was interrupted for up to two weeks, while some ten thousand men worked to repair washouts, replace bridges, and remove thousands of obstructions from the tracks, including entire houses, boats, and even the 190-foot Coast Guard tender *Tulip*, beached on the railroad tracks in New London, Connecticut, for seventeen days. At the Harvard Forest, the needled canopies of white pine caught hurricane-force wind that yanked them to the ground. From that day forward, the forest would never be the same, in either its look or its purpose. More than 2.6 billion board feet of timber in New England was blown down, including most of the merchantable timber at the Harvard Forest.

Photos of storm damage in the Harvard Forest Archives show

trees snapped, tipped, broken, and tumbled. The size of tipped-up root mounds dwarf adult men standing near them in the photos for scale. It is hard to image the power it would take to mangle that much forest. Not just patches but entire landscapes were reduced to a jumble of downed trees. In some of the photos a few lone big trees still stand, the shattered limbs torn from their trunks.

Because of fearmongering that the downed timber posed a fire risk—and to wring what money could still be had from the trees—the Forest, under its then director, Ward Shepard, was, like forests all over the region, put to salvage logging. With nearly three hundred portable and stationary sawmills deployed, photos in the Harvard Forest Archives show tar-paper-shack camps for the crews running mills in the woods, great domes of sawdust, and piles of smoking slash. Crews cut the logs where they lay. With so much downed timber, every pond for miles was used as a storage yard, the water raised with dams to cover the wood. But in one tract—the Harvard tract at the Pisgah forest—it was different. It had always been.

The story goes back to 1927, when Harvard acquired a twenty-acre tract of old growth at Pisgah in southern New Hampshire, regarded as one of the last pieces of primeval forest in the region. Privately owned by the same family for more than one hundred years, it was slated for cutting when Richard Fisher, the first director of the Harvard Forest, spearheaded the effort to save the massive pines and hemlocks three or four feet in diameter at breast height, and more than 140 feet tall. Here stood a "museum of forest antiquity," as Fisher put it, with trees dating to the eighteenth and even seventeenth centuries. His quest became a cause célèbre, capturing the attention of both local and national media. In her story "New England's Virgin Forest in Danger," published in the Sunday magazine of the *New York Times* on July 4, 1926, reporter Mary Lee sounded the alarm:

> The last remaining stretch of primeval forest that covered the New England hills before the pilgrims landed at Plymouth seems doomed to fall under the lumberman's axe next winter unless someone steps forward to save them. The huge survivors of the murmuring pine hemlocks that once sheltered the wigwams of Indians are to be drawn down on the next Winter's snows to the mills along the Ashuelot River in lower New Hampshire to be sawed up into window sash.

Eastern white pine was heavily cut from the start of European settlement. Light but strong, and rising more than eighty feet without a first branch, it was the source of choice for masts, sought by shipbuilders around the world. Aware of the colonists' busy and global masting trade, the Crown assumed exclusive ownership of the best trees. Agents of the Crown were appointed as surveyors to mark any white pine of a diameter of twenty-four inches or larger, measured twelve inches from the ground and found within ten miles of a navigable waterway. Surveyors hacked these prime trees with three hatchet slashes, forming the King's Broad Arrow. A fine of one hundred pounds sterling was decreed against anyone cutting these trees but for the Crown. The British Parliament extended more protections, by 1772 claiming even twelve-inch-diameter white pine. Skirmishes between colonists and British agents seeking to enforce the Crown's fines and prohibition on cutting were called the White Pine War and the Pine Tree Riot—early acts of resistance against British rule. Not for nothing was an eastern white pine depicted on early colonial flags—including the ensign carried into the Battle of Bunker Hill. The colonists wanted those big pines for their own use and profit. In time, nearly all were felled. Yet the big stand at Pisgah, remote, and on rough terrain, still stood spectacularly tall.

Work Fisher's students had done in this old growth at Pisgah was already revealing the dynamic reality of old growth forests. These were not the brooding, decadent, stagnant stands of immense trees people thought. Instead they were far more complex, with the wind, insects, disease, and other forces of nature, combined with the effects of elevation, aspect, and site, creating a dazzling diversity in the composition, density, and form of the trees.

When Fisher—and the *New York Times*—put out the call to save trees of such storied regard, in such a rare example of old-growth forest, the public responded. One ten-year-old boy sent in twenty-five cents; a secretary sent in ten dollars, all she said she could afford, but she offered to type letters at night after work for the campaign if that would help. By winter, Fisher had raised twenty thousand dollars to purchase the forest. Harvard College took ownership with the promise to its benefactors that the forest would be undisturbed by people in perpetuity. But before long, nature had its own way with the forest instead. The 1938 hurricane, just a little over a decade later, laid flat most of the trees at Pisgah. At least Fisher, who died in 1934, didn't have to see it.

The Pisgah's pine giants still tower in sepia images on glass lantern slides. Tucked away in the Harvard Forest Archives in a box, each thick, weighty glass slide is about the size of a baseball card, and individually hand-wrapped in folded white paper. I carefully unwrapped a few and held them up to the light—and was intrigued. Hungry for a better look, I got out an old tabletop projector, plugged it in, and was glad to see its bulb still burned bright. I put in a slide and looked into the past. Here were trees of a size I had not seen in the eastern United States. A man the photographer positioned next to a tree for scale looked like a small, soft vulnerable animal next to a trunk wide as a door. Slide after slide revealed the photographer going for the same effect, with a man standing on a downed tree or just in front of a trunk, each with the same "Can you believe it?" expression echoing down through the years in the image. A few years after those photos were taken, nearly every one of those trees was blown down. But unlike with all the others cut and hauled away for lumber, at Pisgah, nature had its chance to remake the land.

Al Cline was Shepard's successor at the Harvard Forest, after Shepard's departure to Washington, D.C., to run the timbering salvage effort. Cline was no visionary; he himself ran a full-on salvage operation on three tracts of the Harvard Forest at Petersham. But at the Harvard tract at Pisgah, he remained loyal to the promise that it remain a place unmanaged by people. Cline made a controversial decision to let those giant hemlocks and white pines just stay where they lay. It was the beginning of a new day for the Harvard Forest. Founded as a department of Harvard University expected to pay for itself from a sustainable yield of timber, that changed in a single September afternoon, with the loss of two-thirds of the merchantable timber on the Petersham tracts. From that day forward, the Harvard Forest would increasingly be logged only in the service of research. Meanwhile at the Pisgah tract, the lesson those woods have offered ever since is how a forest recovers, without interference or intervention by people. Fisher had seen a living laboratory in Pisgah far more valuable than any timber harvest. Little did he know just how right he would be—even after the hurricane. The tract would turn out to be just as valuable blown down, to research natural recovery after a major disturbance.

Examples of woods left alone to grow and change over time are surprisingly rare. The research generated from the Pisgah tract would become one of the Harvard Forest's signature bodies of work, producing new insights, and rewarding questions decades later. Every five years, researchers would measure and record the condition of each tree growing in this plot. We were headed out to do our part in the effort, in just one of many trips we'd make that fall, counting and measuring the trees.

We wound through New England towns and villages as we made our way to the New Hampshire woods, the Ashuelot River purling over rocks and beneath a red-painted wooden covered bridge. As we drove the two-lane through small town after town, we saw pumpkins on the porches, golds and yellows in the maples, and hay-stuffed scarecrows in the yards. Then Barker Plotkin took an abrupt right, and we left the pavement, potholes jouncing us until the coffee bounced out of our cups. I opened my jaw just a bit to keep from biting my tongue and was relieved when, down miles of back road, we finally parked deep in the woods.

I got out and had a look around. Wild grape climbed the trees, winding its soft yellow leaves into the branches high above us. The fragrance of their fruit was on the air. Barker Plotkin quickly unloaded our gear, and we each shouldered our share. I noticed the old logging road we'd be walking went steeply uphill, out of sight, into the trees. "This is one of the reasons some people won't come here for fieldwork," Barker Plotkin said cheerily, and I shot Saddig a wary look. But this should be no problem, I thought to myself, trusting in my hiking ability and sizing up Barker Plotkin, nearly a head shorter than me. Surely, I figured, I would easily keep up, as she led the way. I knew in minutes just how wrong I was. She is one of those people whose speed defies size. You've known them: the locomotives on hiking trails, insentient to grade. And she kept right on talking, not even winded, as I hoarded my breath. I scrambled in last place, stopping to take off first my Windbreaker, then my down vest. Off came the scarf, then I stuffed my hat deep in my pack. I couldn't figure out how she went so fast, or was it just that she never stopped for a rest? But then, luckily for me, something grabbed her attention, up in a tree. I caught up and saw, looking sleepily down at us, a brown, densely quilled porcupine. It was wedged in a branch where it just met the trunk, all tucked in, and barely awake. It was small and sweet faced, and not moving a bit. Had we awakened it? It looked ready to go back to sleep or to bask in the sun. Could I ever relate.

Barker Plotkin pulled out her map to check our progress, careful not to miss the turn that would take us to the Harvard tract. Marked on the trail with a discrete piece of ribbon down by the forest floor, the tract was intentionally hard to find, and not on public maps. For within this research plot every piece of wood mattered, not only the standing trees, but especially the downed ones: the old-growth logs, the tip-up mounds, each was a clue to the transformation of this forest since it blew down. Barker Plotkin spied the flagging and we headed off the trail, using a compass to help us find the first plot. A small, red-painted piece of rebar, barely sticking out of the soil, eventually told us we had arrived at what we were looking for. It was a corner stake, set by Foster back in 1984. Pisgah was one of his earliest research projects at the Forest. He was only one of many scholars who would return to Pisgah again and again, with a near obsession to probe this rare museum piece of ecology.

We dumped our packs and walked off transects from the first post, searching out the other three pieces of rebar marking the corners. This was harder than it sounds. A lot goes on in the woods in five years—all of it capable of burying a bit of rebar stuck in the ground. Just finding the corners of the plot inevitably was the slowest part of each survey we did that day, as we pawed through leaves and checked and rechecked maps to figure out where the markers should be. After we finally set up the first plot, we divided up the tasks. I would be the recorder, writing down the species measured, and each tree's condition. Saddig and Barker Plotkin would work their way over the forest with a tape that measures the diameter of trees at breast height, calling out the species and dimension of every standing tree in the plot, chalking off each trunk as they went, to avoid double counting. Recording their count let me be useful, while still having plenty of time to enjoy these woods, which looked like none other I had seen, anywhere. It was a landscape without edges, and full of surprises. I sat on a downed log two feet across, finding it soft as a sofa. Though nice at first, it started crumbling beneath me, sending me bolting to my feet. Everything around me was softened by moss and rot and time, logs melting into the ground.

Barker Plotkin, taking a step, sank to her calf in a compost of leaves and downed wood. It was my first horizontal forest, a temple of dead wood. It was impossible to walk anywhere in a straight line because of all the downed trees, their girth so big that many required a clumsy clamber around or over them. This enforced the pure physicality of the place; this was not somewhere to be just walked through, without notice; every step had to be considered. Was this log—solid? This ground? Or would my foot plunge into another dimension, the whole surface on which I would sit give way? That had already happened once, as this forest elder made its own rules for visitors to its cool and padded realm.

In this place of preternatural quiet, time seemed to slow down. Everything was bigger and older than us, turning on a great, slow wheel of deep time. Neil Pederson, a senior ecologist at the Harvard Forest and a specialist in dendrochronology—the reading of the cores of trees—had dated one tree in the tract to 1674. The cushioned ground underfoot hinted at an unseen world belowground, too: the reaching tendrils of mycelia, billions of busy bacteria, and every tiny, crawling thing devouring these fallen monarchs, their carbon riches mined by resurgent new life. For these trees' role in the ecology of this forest was just getting started; 1938 is yesterday in forest time. Deadwood isn't dead at all. It's just finding its new purpose, as the food and moisture source that nourishes the next phase of life in these woods.

"I've got hemlock," Saddig called out, as I struggled to keep up with the cadence of their work. "Beech," Barker Plotkin called. Hemlock, birch, beech: it would be the roll call of these woods. Once a stand of massive white pine, today the Harvard tract has remade itself, and its biodiversity is exploding. Logs were quilted with moss and fungi. Hemlock, beech, and black, yellow, and paper birch trees rooted avidly in an all-you-can-eat cafeteria of woody debris. Trees long suppressed in the shade of the now-fallen giants, or seeds germinated after the hurricane, had in the more than seventy years since soared into a new canopy. Birch roots gripped the tipped-up root wads of the fallen white pine, the tip-up mounds of the prostrate giants providing a perfect platform for new life. Yet we found no white pine in the regrowing forest; today this wood is dominated by hemlock.

"Pine, which once reigned here as the largest and most visible species, has disappeared," Foster wrote in his essay on the Harvard tract at Pisgah in *Northern Woodlands* magazine.

Natural processes alone caused this remarkable change, confirming how kaleidoscopic our forests really can be . . . The dramatic shift in composition has major consequences for the ecosystem. White pine grows faster and larger than most other species, and its longevity allows it to reach super-lative heights and girth. This meant that Fisher's old-growth stands at Pisgah had among the greatest amounts of standing timber of any forest in New England. In contrast, the new stand—with its much smaller and slower-growing hemlock and beech and much shorter-lived red maple—will never support such massive trees or staggering amounts of wood. The rapidly re-growing forest will become increas-ingly impressive and may one day assume old-growth quali-ties, but it will never be the one that Fisher saved.

This transformation to a new suite of species was one surprise, but Pisgah had another lesson. Leaving the forest alone had allowed it to naturally regenerate more robust and diverse growth, without the air and water pollution, road construction, risk of fire from slash burning, and soil compaction and destruction of bio-diversity that comes with salvage logging. Yet another discovery for scientists has been the degree to which the disruption of the 1938 hurricane was still shaping this forest more than seven decades later. It left a legacy of complexity that is still unfolding, in a forest diverse in structure and form, and an understory of still slowly decomposing old growth. Finally sitting firmly—on a rock—with my clipboard, I felt I could hear the hyphae growing in the ground, and the mushrooms, lichens, mosses, and molds burgeoning on rocks, stumps, logs, and trunks. The sky was bright between the new growth of trees reaching for the sun. But we worked far below, in a dim gloom of the hemlocks and beeches. It was silent and still down here, like working in a vast, underwater shipwreck.

These trees were a look into a procession of time revealed at a magisterially slow tempo. Logs big as beached whales reset the imagination to a new scale of what trees can do, given enough time. But on our way out of the forest came the biggest reset in what I thought I knew about these woods. Suddenly, there stood a single old-growth white pine, somehow left untouched by the storm, supreme above all else in its grove. A scatter of cinnamon needles was everywhere under it, like a tribute to a queen. What I had thought of as mature trees that I had been under and amid all day suddenly seemed mere sprouts, topping out not even midway up the skyscraper of this tree, tall enough to carve the high winds. It was a recalibration of what wonder is, a hint of what this forest must have been like—and the power it took to topple most of it. As I looked at the giant trees near it, laid over on their sides, I felt I could hear the wind howl.

Other trees at Pisgah also escaped the storm. Later that fall, on another field trip, we went to go see them. North Round Pond, as the tract is called, had seen some cutting, but it also included some big old trees spared by a roll of the land that sheltered them from the hurricane winds. On a perfect fall day, Pederson led David Orwig and a band of research assistants on a coring expedition to the grove. Pederson wanted to learn the stories of these trees—how old they were, for one, and what happened to them after the storm roared through. For if you know how to hear their stories, trees, as Pederson likes to say, are quite talkative indeed.

Pederson brought a questing curiosity to his investigations, a lively stream always seeking bigger water. His high school thirty years ago had no environmental science classes, and Pederson was one of only two math majors in his junior college. But a class he took in science and society opened his eyes. "I thought I was going to be a math teacher and a lacrosse coach," Pederson said. "Then I was learning how cool forest ecosystems are." He has since crisscrossed the eastern United States and remote reaches of Mongolia, and even the ground on which the World Trade Center once stood to core old ships, ancient deadwood, and trees of all sorts to learn their history. His fresh questions led to startling work that made headlines around the world. Pederson and his collaborators aged a ship timber found at the World Trade Center site (1773) and hypothesized a likely contributor to Genghis Khan's success: rain that stoked a bumper crop of steppe grasses for the war machines of his age—horses.

Pederson had a gentleness and an old-fashioned loyalty about him. The death of a mentor visibly saddened him for days. When it came his turn for the weekly lab presentation at the Forest about ongoing research, Pederson stuck a pen in his mouth and let his graduate student do the talking, to gain the experience. I could tell the seasons by the length of his beard.

"It sounds silly, but don't kick or step on the downed wood," Pederson said, as we entered the grove. That's because he intended to core not only the standing trees, but also every piece he could discern on the ground. "We care about everything at this site," Pederson said. All around us were trees the likes of which I hadn't before seen; even their bark was different. Red oak gets a different look at great age: "It's nothing like classic red oak— look at that smoothness," Pederson said. "It gets that powdered-sugar look."

Trees like these were important to core because of what they could teach about how fast and in what ways they were growing, and the stand structure, or mixed ages and types of the trees. The woods, so quiet when we arrived, were soon loud with the sound of scientists coring many different trees at once, in a chorus of squeaks and squeals as the trees' grain gripped corers pressed, turn by turn, to their heartwood. The researchers worked their way across the plot, hanging a temporary metal tag on each tree now in the service of science, measured, mapped, numbered, and cored. "What do you think, maybe 1780?" Pederson said to Orwig, estimating a tree's age as researcher Brian Hall pulled the core of a red oak bigger and older than anything else I had seen in these woods before, and more than twice the age of the big oak back at Petersham. The metal of the corer, by the time Hall pulled it free from tree's heart, was hot to the touch from the friction of the tree. "They don't like to give up their secrets easily," Orwig said as he cored into a neighboring tree. "We don't call them hardwoods for nothing."

A black birch one tree over released its wintergreen scent as a scientist slid its core into a straw for safekeeping. Big or small, young or old, every species and each tree in the grove had its turn as the scientists kept at it, coring about a hundred trees in just this one plot. The crickets trilled, the trees squeaked and groaned, and scientists called out tag numbers and tree measurements from three directions at once. They were just getting started and still had all the downed wood to go. I didn't expect anyone back at Shaler Hall until dark.

A few weeks before, Pederson and I had talked about what it means to know a tree's story. Change is the norm in a forest, but it happens—usually—more slowly than humans can perceive. Trees reveal history though, not only of their own lives, but the forest around them, in their wood. Not the light springwood, that's an echo of the energy stored in their roots from the year before. But the darker band of summerwood a tree makes from its environment every year—that's the storyteller ring. "Trees are different from animals, they can't run from fire or drought, they have to sit there and take it," Pederson said. "They are witnesses to history, and they take the abuses of time, they internalize it, and that helps us learn so many things from them that we don't expect." His work is old-style natural history, done out in the field, and for months in the lab, sanding tree cores, and pouring over them under the microscope to read the memory of trees. The density of rings and their variation, the scars and signs of suppression of growth, all have something to say. "It's beautiful to me because all the information is in there," Pederson said.

A skeptic by nature, he'd learned to cross-date trees, relying not on one tree but a grove to tell the environmental history of a place. Trees can be tricky. Some, when stressed, won't produce a ring. Pederson calls them zombie trees, hiding many years of their actual time on earth. A big tree is not necessarily old, nor a small one young. A big tree may just have lots of sun and water. And many are the crabbed, tormented trees of modest size, still waiting for their chance to grow, even over many centuries not becoming large, yet persisting to great age. "The old ones often are not the ones we think they are," Pederson said. "But one of the beauties of our science is it is precise. Unlike ice cores, or sediment records, if we do our work correctly, the error bar is . . . zero."

Steeped in the historical perspective of forests and trees, Pederson has a geologist's mind-set that helps him take a step back, be data driven in his conclusions, and even just keep an open mind. "I get frustrated listening at conferences about extinction," Pederson said. "I have faith in genetics. Trees can leap over cities. People say we are going to lose all these species. Well, maybe. But maybe they are going to adapt and evolve. Evolution isn't in one direction." There is an inherent hopefulness to the long-term view, Pederson said. "When I think from a geologic perspective, and this new infancy of genetics, we know now that plants have multiple copies of genes; maybe they are adapted for these extreme changes. And they throw a ton of seeds. It may not be this landscape that we are used to in the future, but it will be a landscape."

After that trip to Pisgah, and those conversations, I thought about the big oak in a new way. What had the tree already been through, and how had it fared? How had it managed the vicissitudes of its changing environment? The big, fat saucy rings in its core, year after year, seemed to depict a life of ease and smooth sailing. So I made another trip to the Forest archives to see what all had gone on in its life, and in particular what had happened when the hurricane raked over its little part of the world. As the Christmas holidays approached, I saw the perfect opportunity for a deep archive dive.

Shaler Hall emptied out as most of the staff left for a holiday closure. The heat in the offices was turned down to sixty, and just a few hardy scientists came and went; mostly, the place was deserted. Even the animals in the Forest were hibernating. But I knew where it was warm, bright, and cozy: in the archives, kept at a toasty constant seventy degrees for the sake of the records there. So one deep winter night at the stub end of the year, I packed a velvet pillow, pencil box, and a bunch of notebooks in my bag, went over to Shaler Hall to slip the key to the archives building off its hook, and stepped out in the night. My feet made tracks in the falling snow as I crossed the courtyard to the low brick building where the institutional heart and memory of the Forest abides.

As a writer and reporter, I've worked in all kinds of record repositories. Local, state, and tribal government offices, state capitols, the Library of Congress, local historical societies, grand public and private and university libraries, athenaeums and special collections of all sorts. State archives, courthouses, and judges' chambers. But it's not often I've found the equal of the Harvard Forest Archives. Here were polished long cherry tables lined with Windsor chairs, as if for a banquet—and indeed that is just what awaited me: a feast of resources. Here were more than a century's original handwritten records of forest management, scrupulously organized. Hand-drawn and colored maps on velum and paper, stored in oversize file drawers with a delicious smooth slide. There were original historical photographs, glass lantern slides, and even a three-dimensional wooden map with pushpins marking the locations of salvage logging operations after the 1938 hurricane. There were newspaper articles and government reports in their originals—what a luxury to read on paper these days—and probably a hundred linear feet of bound master's and doctoral theses. There was Sarah Mann's diary, and even the original account book of the Sanderson farm, begun in 1776. Its leather binding was worn to suede soft as an earlobe, and the pages of thick, creamy paper had aged to sepia tones. The family wrote upside down on the backs and in the margins of pages of prior years, to save paper as time marched forward in a record of most every expense and sale on the farm, kept through three generations.

In an age when many institutions don't provide much space, staff, or budget for the preservation of libraries and archives, let alone memorabilia, here was the first Harvard Forest director Richard Fisher's Harvard cap and gown, carefully archived in a box up on a shelf, with a pair of white cotton gloves tucked inside to handle it. A stuffed bobcat named Bob lorded over it all from atop a file cabinet in the center of one of three reading rooms, with a little hand-lettered sign reading PLEASE DON'T FEED THE BOBCAT.

I settled my pillow on the elegant but fiercely firm wooden chair at the table under Bob and got to work, enjoying yet another great gift of this place: the expertise of Elaine Doughty, the forest's archivist. She had created an index of materials she thought I might enjoy and find helpful, and I was eager to start working my way through it. There was no record of this particular oak. But she suggested that a read through the stand records of the area, or compartment, as it is called in forester's jargon, of the Forest where it grew would give a hint to the tree's life and times. So as the snow sifted past the windows, I dug in for a nice long winter night's read.

Maps showed me that the oak grew up by the stone wall in a former rough, unimproved pasture, with a cultivated field across the other side of the wall. The soils in its root zone were classified Acton stony loam and Gloucester stony loam; the stone wall could have acted as a practical dump for rocks gleaned year in, year out from the ground. The soils were said to be moderately or poorly drained—and I believe it, because of the seasonal stream I would see passing just beyond the tree in the early spring, and the persistent dampness in the ground to the east of its roots. The call of spring frogs alerted me to the vernal pool I discovered one spring, following the sound to its source, just one hundred strides north and west from the big oak. A map in 1908 described the parcel where the tree grows as "poor hardwood," so by then this ground was growing up in trees. Trees in the oak's parcel were estimated in 1908 to be two years old.

By 1913, the stand was described on another map as hardwood-cordwood, and the area across from it on the other side of the wall was still open. By 1919 both sides of the wall are described as inferior hardwoods, and by 1923 the area was dominated by gray birch. By 1937 the woods on both sides of the wall have grown to pine and cordwood, which explains why today I still see big pines in the forest there. By 1947 the stand on the oak's side of the wall is given to red oak and red maple, and by 1986 trembling aspen makes an appearance nearby, a tree I don't see there today.

In addition to the tree's probable history, the stand records of damage typically sustained by young trees also gave me some sense of how remarkable it is that the oak not only germinated, but survived. Terminal buds at the central leader of its growth could have been nipped by birds, or its tender growing bark whipped through by neighboring branches. Sunscald, insect attack, and girdling by rodents were all possibilities. And then there was all the human management going on all around it.

Crews were put to work cutting trees sometimes just to open views from the road, or to give them work while snowbound. There was a fire nearby, and many episodes of "liberation cutting," intended to generate big trees. Poles and posts were harvested, cordwood was cut, and stands were weeded, thinned, harvested, clear-cut, and girdled. A severe ice storm in the winter of 1921–22 splintered the canopies of trees and left the survivors suffering from top rot. Yet the little red oak by the stone wall was each time spared. Then came the September 1938 hurricane, and the damage—all around it—was near total. "Damage was severe and complete, the ground was left encumbered with a tangled mass of down trees, lying in a general northwest direction," a note in the archive as to the fate of Compartment III reads:

> Some scattered individuals along the western edge remained standing, where resistance was low, and wind able to break through. Owing to tropical rains preceding and accompanying the storm, less than normal anchorage was offered by the sodden ground, and wide, spreading, shallow root systems were quickly loosened by violent and erratic gusts of wind attending the hurricane. Everything was cut and bucked into logs and sold where they lay, the slash piled and burned while snow covered the ground.

Checking a map of hurricane damage, it's clear that parcels right next to the oak suffered severe damage—just across the wall, the map marks the area "destroyed." Yet right where it grew, damage was rated as "moderate," with only 11 to 25 percent damage, by some quirk of topography, or chance. One advantage, for sure, was the tree's young age at the time of the hurricane: it was just thirtysome years old, and probably able to just bend down with the wind. Or perhaps its roots, hunkered by that stone wall, stood a better chance of not ripping free from the ground. I've often noticed a big rock gripped by the oak's roots at its trunk, too, and wondered how big that rock is under the ground. And perhaps this is how the tree's lean and twist got started?

Another big ice storm struck in 1962, and a devastating gypsy moth attack knocked back the oak's growth not once, but twice, in 1944 and 1981: I can see that in the rings in its core. Its summer-wood both years has a distinct pinch, one of the few years in which the tree noticeably slows its growth. Indeed in this oak, as in other red oaks at the Harvard Forest, the last twenty years show one dominant characteristic: growth, the fastest ever recorded, and faster than in any other species.

"Trees are like people," Pederson said. "They are individuals. But as a population, they say something. In the late wood, if the rings are getting bigger, that could be climate change. If we are seeing it at a larger scale here, and other places, then it's less likely to be local ecological conditions." Trees tell us history, Pederson said, but they also hint at the future. Trees are talkative. Our job now is to hear them.

CHAPTER SIX

THE LANGUAGE
OF LEAVES

THE FIRST TIME I went to the Harvard Forest to visit John O'Keefe, my walking companion who helped me pick the oak, I didn't know exactly what I would find. I had heard about him and his work through the Richardson Lab and mutual friends and wanted to come learn more about his study of seasonal changes at the Harvard Forest.

If the mark of a lifelong New Englander is a penchant for hard chairs, and a car with more than three hundred thousand miles on it driven with skinflint pride, O'Keefe fit the bill. His metal desk was tucked against a bookshelf to one side of a room packed with field guides, and a wall of metal file cabinets full of his papers and records lined the other. His battered briefcase was slumped on a wooden chair, with a small braided rug for a pad. He had something of the look of an eaglet about him, with his white, balding pate and steady, observant eye. Cool in the summer and snug in winter, his office was the hidey-hole of an adaptable species, using well-thumbed field guides on wildflowers, weeds, trees, and shrubs for nesting material.

O'Keefe had made his way to the Harvard Forest after volunteer service in the Peace Corps in Africa, and seven years of flying F-106s for the Massachusetts Air National Guard during the Vietnam War era. The F-106 was a specialized missile-equipped aircraft, designed to shoot down bombers, and was the primary all-weather interceptor aircraft of the U.S. Air Force from the 1960s to the 1980s. As I got to know him, I better appreciated the surprise that this pie-baking, folk-music-loving, sunny soul had trained as a jet fighter pilot, groomed for aggressive aerial combat on the "Ultimate Interceptor," the Convair Delta Dart. "It was my way of dealing with the military," O'Keefe said.

After flying over miles of forest all those years, O'Keefe found himself, after he was discharged, wanting to know just what was down there, and how it all worked. His undergraduate degree from Harvard College was in sociology, but O'Keefe was always interested in science. So he set himself to earning fifty credits in biology at the Harvard Extension School and, at age forty-two, graduated from the University of Massachusetts at Amherst with a Ph.D. in forest ecology. When, in 1988, John Torrey, then director of the Harvard Forest, was looking to develop the Forest's museum as a centerpiece of public education on the Forest's land-use history and ecology, O'Keefe came on board as museum director. He retired from that job in 2009, but never left the place, keeping an office as an emeritus researcher at Shaler Hall.

At the University of Massachusetts, one of his advisers, Bill Patterson, had got O'Keefe interested in phenology, the age-old practice of observing and recording the timing of seasonal events in nature. O'Keefe had helped Patterson develop a trail for making systematic phenological observations at the University of Massachusetts forests and helped Patterson with his research. When O'Keefe came to the Forest, he took Patterson's advice to create a trail at the Harvard Forest that was close by and simple, so surveys could be repeated easily and often. Thus began O'Keefe's phenology trail out the back door of Shaler Hall, which he could do on a long lunch break.

O'Keefe started off with more trees, but to keep it manageable eventually cut back to about fifty trees on a two-and-half-mile loop through the Forest. By the time I caught up with him, O'Keefe was starting his twenty-fifth year of surveys. He was observing in the leaves of the trees how the timing of the seasons progressed, relative to years past. He measured little, but for the occasional snow depth, or length of an unfurling leaf. But what he did do was look closely at a set number of tracked trees, chosen to represent a

range of species, heights in the canopy, and forest environments—
dry, wet, open, and shaded. He also made systematic notes of his
observations on data sheets he created for the purpose, filled out
the same way each week, year by year. Without initially intending
to, he found that over the years he had created an important
record, ripe for mining patterns in the data. Andrew Richardson
was among the first of many researchers to use O'Keefe's records
in a series of influential scientific papers about the effects of
climate change on forests.

I knew of O'Keefe's data, but had not had a chance to spend
time with it. So one night during my first year visiting the Forest,
I asked O'Keefe to leave his door open for me when he left for the
day, and for permission to sit at his desk and read through his
files. It was a cold January night, with the moon riding high, and
I didn't even need a flashlight when I headed over to Shaler Hall
after dinner. I clicked open the lock on the big heavy back door,
letting myself into the dim hallway lit only with the red glow of the
exit sign over the door, and headed down to the basement. I
flooded O'Keefe's office with his overhead fluorescent lights and
got to work.

When I pulled the first few files to sample them, I was mostly
just hoping I would be able to read O'Keefe's handwriting and
make out his abbreviations. But what was revealed as I kept at it,
sitting at his big metal battleship of a desk, was not only a diary of
the seasonal progression of the trees. In the notes written in the
margins of his data sheets was also an intimate look into the life
of the Forest, as it wheeled through the seasonal cycles of the year.

The week-by-week portrait was made with all five of his senses: how firm the tree's buds were, or whether they had softened and were getting ready to crack. The sound of the first call of wood frogs, the scent of mineral soil as the frost melted from the ground. The sight of the leaves' first emergence; the filling and draining of puddles; flow of the streams and first unfolding of woodland flowers. The autumn colors of the leaves, the thunk of acorns; frost flowers and ice on the puddles, and the wintergreen taste of birch bark. Here was a place richly and closely observed, right down to the mud and blackflies. With nothing more than a pair of binoculars, six-inch ruler, pencil, and clipboard, O'Keefe by walking the forest again and again had amassed a detailed calendar of the seasonal year, his tiny handwriting in No. 2.5 pencil recording local events with planetary implications. I knew then what I wanted: to start joining him on his walks. I was thrilled when he agreed.

We would walk O'Keefe's survey loop together for the better part of two years. Yet we never saw the same thing twice. Nature is always becoming. Rather than limiting our exploration and discovery, walking the same path was actually a way to see more, not less. I had my own landmarks on our loop—favorite features, experiences, trees, ferns, plants, sounds—and they changed through the seasons. I could only imagine the dimensions and layers this walk must hold for O'Keefe. It was the synchrony of the forest, with every plant and animal and insect in its place, in its time, for a reason, in concert with all the other plants and animals that to me held such beauty. The coming of spring was like the beauty of an orchestral fugue, with each instrument joining on cue to build a symphonic composition.

We started our walks for the seasonal year in April, the back door of Shaler Hall banging shut behind us as we headed to the dirt road that would lead us to O'Keefe's loop. He had his clipboard, pencil, and binoculars; I had my camera and notebook. He had shed his wool hat and boots, and I was wearing only one jacket and no wool scarf: sure signs of spring. The blackflies were not yet out, nor the mosquitoes, but the snow was off the ground. It was mud season.

He tipped his head back to look more closely at a sugar maple, using binoculars to examine its more than two-hundred-year-old canopy. "They are just about to open," O'Keefe said, studying the tree's swelling buds. It would be a six-story-high tower of lime-green new leaves pretty soon, then radiant gold in the fall. For sheer drama, this tree was one of my favorites on the walk because of its size and age. It was also right by the road. That meant you could see it entirely from at least two sides. We walked on to the striped maple that so entranced us both, to find its buds swollen, but not yet split and spilling out the pretty tresses of bloom that festooned it in later May. The forest was so on the verge, about to inhale its first big breath of the year.

When and how leaves developed through spring and fall was the main subject of his work, and he tracked his study trees in weekly, even biweekly observations, from the first swelling of buds to emergence and every stage of development. He watched for color, texture, size, and shape, observing the trees' incremental changes until each tree he studied had all its leaves. He resumed his observations in the fall, from the first coloring of the leaves until 95 percent of the trees he studied were bare.

Trees both affect and are affected by climate change in ways that are measurable—and important. Leaves are wonderfully legible indicators that reveal the workings of the forest, and influence of the environment. Budburst is sensitive to temperature, and the warmer it is, the earlier it comes. Once trees begin photosynthesizing, they also gobble carbon dioxide. As the temperate forests of the northern hemisphere green up, they have an atmospheric impact measureable on a local and planetary scale.

O'Keefe's records show how climate change is altering the seasonal timing of the forest. Leaf-out is coming earlier, with an advance of spring by nearly five days on average since O'Keefe started making observations more than twenty-five years ago. The onset of the first frost in the fall had changed even more. O'Keefe used to observe hard frost as early as the first weeks of September. By the time we began our collaboration in 2013, the first frosts often weren't coming until late October. On average, spring is coming earlier. Fall is coming later. And winter is being squeezed on both ends. Everything in the woods reflected these changes, from the level of water in the vernal pools and springs, to when the blackflies were biting, the ground frozen, or leaves budding out or finally coming off the trees. It wasn't a matter of conjecture or political argument; the discussions of who does and doesn't "believe" in climate change, in editorial pages, news reports, and congressional debates, frame this all wrong.

Climate change, the trees, streams, and puddles, and birds, bugs, and frogs, attest, is not a matter of opinion or belief. It is an observable fact. Leaves don't lie; frost isn't running for office, frogs don't fund-raise, pollinators don't put out press releases. What O'Keefe had compiled, while taking all those walks, was the testimony of an unimpeachable witness: the natural world. Including the big oak, my witness tree. O'Keefe's survey walks were a way to observe the planetary dynamic of climate change on an intimate scale and see its effects, tree by tree. It was like freezing the action of a movie, to study one frame.

Systematic observation was important. Emergence of leaves is a delicate business. Trees in a sunny spot will leaf out before the same species in a hollow of shade that collects cold air. Even the same species of tree in one general area could leaf out at very different times—perhaps their rooting was different, or the below-ground environment varied. I saw this with the big oak, which came into leaf later and shed its leaves sooner than another red oak right next to it.

Each species took its turn, and even each tree, breaking its buds in stages of swelling, cracking, emergence, and leaf growth. What a drama it was, impossible to see over the course of a day, but vividly apparent over the course of a week. The transformations came so fast we sometimes walked not once but twice in a week, so as not to miss too much. As the great naturalist Aldo Leopold wrote in *Sand County Almanac*, "In June as many as a dozen species may burst their buds on a single day. No man can heed all of these anniversaries; no man can ignore all of them."

O'Keefe's walk took him through a mostly deciduous wood, with several streams he observed; a hemlock grove with a spring and a vernal pool; a black gum swamp cushioned with ferns and moss; the site of an old burn grown back with paper birch; and finally up behind the Barn Tower, with its cameras and other instruments, where the big oak grows. Old stone walls and dirt roads everywhere reminded us of the past. Scientific equipment from laundry baskets collecting leaf litter, to a so-called megaplot in the Forest, where every stem bigger than a pencil is tagged, marked, and mapped, reminded us of the Forest's present purpose as a living laboratory, with hundreds of research experiments under way.

We checked maples and ash and beech, then made the left turn in our path that turned us from birdsong and sun into the cool, quiet glade of the hemlock grove. Even with a blindfold on, I would have known we were amid the hemlocks now, so different was the feel of the ground underfoot, with the layers of needles piled thick and soft as a mattress. The light was suddenly filtered, too, and the breeze stilled by the hemlocks' dense cover. The temperature was at least five degrees cooler. A wet seep in the woods that we watched was tip-top full; we noted its flow, visible just at the surface, before the water dove back underground. The water was high enough now that we crossed over its bright shine by stepping on mossy rocks. I listened hard for frogs as we started getting close a little farther on to the vernal pool. As soon as the first open water had showed at its edges of melting ice just a few weeks earlier, the wood frogs had found it and begun calling for mates. Wood frogs don't make the classic high-trilling sound of spring peepers. Wood frogs' odd quacking call was one of the earliest signs of spring in the woods, and we could hear them from a long distance away.

Wood frogs make quick work of their mating time in the gelid water. Now the pond was still once more, and their eggs, already laid on subaquatic vegetation, were maturing. Wood frogs return each year to this same pool of their birth. This pool was seasonal, drying up entirely most summers as the trees drew down the water, after the young had hatched and metamorphosed from free-swimming, vegetarian tadpoles to breathing, walking, carnivorous young and the frogs had migrated back to the woods. The water regime of the pool was critical to their survival. Too much water, and the pool would sustain fish that gobble tadpoles. Dry up too soon, and they would not have adequate time to grow the legs and lungs they need to survive. Their lives hang in a delicate balance of conditions in a short window of seasonal time.

We turned left again and walked through the verdant glade of ferns and moss in the black gum swamp. As we left it, I was stunned by a male scarlet tanager, a neotropical migrant bird. Big and impossibly bright, it sprang from its branch and flew, a vivid vision of red with black wings. It had arrived here likely all the way from South America, for the feast of bugs just starting to hatch out in these woods in time for it to feed its nestlings. It wasn't just the presence of this insectivore's food supply that was critical to its nesting success, but the timing of when it was available. I saw a female tanager, with its demur olive-yellow plumage, near the big oak the same week. Could this be the other half of the pair?

We crossed into the old burn, where the sun streamed through the white birches not yet in leaf, warming a thick brown litter of leaves from last year. O'Keefe waved me over to have a look. There they were, the belles of the spring wood: pink lady slippers. Some were just then lifting their flower buds from tightly crooked stems; others were beginning to color; while yet other plants in the sun were already fully in bloom, the pink purse of their petals opening to their inner recesses, inviting a bee. One of the Forest's showiest native orchids, stands of them were everywhere as we walked, plump, pink, pendant, and spectacular. Only during this six-week window, between their first peeking up from the bare ground in spring, until mid-June, when the leaves would once again cover the trees, would the sun beam down on their understory niche as it did right now.

This was the splendid time of the spring ephemerals, the woodland wildflowers precisely timed. They had been coming on with the sun since the spring equinox, the display growing in color and diversity as the sun gained in its height and heat, day by day, capturing the brightest light the understory would receive all year, right between the melt of the snow and the leaf-out of the trees. Four-petaled bluets, their blooms the size of a pinkie nail, paved the crown of the wagon road in nodding blue and white flowers. Sun flecks found pools of deep purple: the broad-leaved wood violets, turning their white-whiskered faces to the sun. The elegant, single nodding bloom of sessile bellwort, with its graceful, winged leaves, stood in creamy-white perfection at the feet of the trees. Bees, wasps, flies, and beetles sought these early-spring nectar sources, returning the favor with pollination, in a meetup essential to each.

The synchrony of all the lives in this wood—the flowers, the trees, the bugs, and the birds—wasn't just beautiful, it was functional, essential to food, shelter, nesting, and procreation. It was the miracle of spring we see in the temperate zones every year, a resurgence of the green plants that make nearly all life on the earth possible, noted by people everywhere, always—until many stopped paying much attention at all. By the twentieth century, even the value of phenology as a discipline came to be dismissed as an anachronism, a hobbyist's holdover from the nineteenth century. It is still today regarded by some scientists as a squishy business. "There's recognition of the value," O'Keefe said. "But it's still thought by many to be natural history, not science."

We arrived at the Barn Tower and began exploring O'Keefe's study trees, more than twenty in this part of his survey walk, including the big oak. I'd been watching for its first leaves for weeks. While the black cherry and red maple were already getting a start on the season, the red oak's buds remained closed. But the base of its trunk was disappearing in a rising tide of green. The tender new growth of ferns softened the stern winter face of the oak's bark. Everywhere around us as we were walking on a birdsong-bright spring day, we saw buds of ferns in tight spirals. They were just starting to uncoil and stand taller, opening their fronds wider by the day, like sleepers stretching their arms and raising their heads after a long winter's nap. Never among the earliest trees to leaf out, the red oak this year was even more delayed. It was a particularly cold, late spring. Yet the spring just a few years before in 2010 had been the warmest and earliest on record, and 2012 the runner-up. This sort of year-by-year inconsistency is exactly what helps confound the climate-change conversation. It is the long-term record—climate, not weather—that counts. That's where the value of a record like O'Keefe's comes in. Put together with climate data and other analysis, O'Keefe's long-term record makes it possible for scientists to investigate connections between the changing climate, the timing of spring and fall, and the ecological function of the Forest.

Discerning the workings of the natural world in seasonal timing has a long history. Dr. Jacob Bigelow, a Harvard University professor, in 1718 published a paper on blooming peach trees in southern and northern locations in Canada. This stimulated more studies by botanists in Europe, leading to the creation of the odd term *phenology*. The roots of the word are the Greek words *phaino*, meaning "to show" or "appear," and *logos*, "to study." It's from *phaino*, too, that we get *phenomenon*, and traditionally, phenology has consisted of the study of the timing of biological phenomena in nature and their relationship with the earth's environment, particularly the climate. The Belgian botanist Charles Morren argued that like meteorology, botany, zoology, physiology, and anthropology, this merited being a scientific discipline unto itself: phenology. He is credited with first use of the term at a public lecture at the Belgian Royal Academy of Sciences at Brussels, in 1849.

Phenology's roots are in old-style, hands-on observation such as O'Keefe's, practiced long before the term *phenology* was invented. The longest continuous written phenological record is probably of the first flowering of cherry trees at the royal court of Kyoto, Japan, dating back to A.D. 705. In Europe, French records of grape-harvest dates in Burgundy stretch back to 1370 and have been used by scientists to reconstruct spring-summer temperatures back into the Middle Ages.

In Britain, Robert Marsham began recording what he called his "Twenty One Indications of Spring" in 1736 on his country estate in Norfolk, England. He tracked the seasonal stirrings of animal life—croaking frogs and toads; singing nightjars; pigeons and nightingales; arriving swallows and cuckoos; rooks building nests—and all manner of plant activity, from flowering snowdrops, wood anemones, and hawthorns, to leafing birches, elms, oaks, beeches, and horse chestnuts. The recording duty passed from one generation of Marsham's descendants to another until the death of Mary Marsham, in 1958.

Britain's Royal Meteorological Society also published the findings of a network of as many as six hundred local observers from 1891 until 1948, managing even during the war years to marshal recorders in districts throughout the United Kingdom. Their work was organized in hand-drawn tables reproduced in foldout charts in the society's journals, packed with detail. As recently as the 1940s, it seems it must have been expected that all sorts of people from across Britain—not only scientists or specialists—had the knowledge and interest to recognize and track a wide variety of species at many life stages. Recorders logged the

comings and goings of bugs, flowers, leaves, and birds; the call-
ings of toads and frogs; as well as indigenous and migratory
insects in both first hatch and second. They tracked ninety varieties of plants, from alder to yew, from first flower, to first leaf, full
leaf, ripe fruit, full tint, and leaf fall. They listened for the song of
the cuckoo, thrush, and lark and watched for the building of rooks'
nests, the appearance of fledglings, and the last departing swift.
These were often just ordinary people, paying attention to their
world.

In addition to data recorded in tables, observers contributed
narrative reports. Their writings depicted a keen attention to the
wonder and beauty and sheer dynamism of the natural world right
around them, richly woven into their daily life: "Many gulls driven
inland by storms. Thousands of birch seedlings in gravel where
siskins flock, feeding in early spring. Chaffinches by hundreds in
the beech mast. Winter wheat four inches high. Good lamb
pasture in March. Autumn tints by September 10. Squirrels
remarkably scarce," one observer wrote in 1925. Not only did
people think such observations were important enough to record,
but that they were important enough to publish in fine hardcover,
gold-stamped volumes that still abide in major libraries. How
interests change. At Hayden Library at MIT, the most recent due
date on one of these carefully produced, beautifully bound books
when I borrowed it in 2014 was 1955. The hand-stamped date on
a paper card was still tucked in a manila pocket, pasted on the back
book cover. Reading that due date, and the book, I was struck by
the sadness of this roll call of the lost. Not only had many of the

species recorded in 1925 since become locally rare or extinct, but even the habit of paying close attention to the natural world, and the ability to discern and understand what was happening in nature through the seasons of the year, had mostly passed from common knowledge. Not only in Europe, but even more so in the United States, despite our once diverse and robust phenological traditions.

As in Britain, in America lots of people once paid closer attention to nature's calendar. Long before he was the nation's third president, Thomas Jefferson was a devoted phenologist. Jefferson kept copious garden notes beginning in 1766 and continued his journaling for more than fifty years, illustrated with hand-drawn maps of his planting beds. Even while embroiled in matters of state, his dispatches home to Monticello inquired incessantly as to the progress of his vegetables, what was blooming in the garden, the arrival of the mockingbirds and their spring hatchlings.

In 1852, the Smithsonian Institution appealed to the American public to gather copious notes on meteorology and a "registry of periodical phenomena." Thrice daily recording of weather observations was required, as well as the ability to appreciate three kinds of dew and frost, multiple varieties of fog (falling, rising, dense, and slight), and innumerable clouds, skies, and rain (continued, interrupted, shower, general, partial, drizzling, slight, moderate, heavy, or violent), not to mention snow, sleet, and clear frozen raindrops versus the white ice of hail. Phenological observations were to include leafing and flowering, fruiting, and leaf fall of hundreds of species of plants, as well as the "first appearances, cries, and general peculiarities of habits" of toads, frogs, turtles, lizards, and snakes; the first appearances and spawning of eel, sturgeon, herring, shad, and salmon, and first appearance and cries of insects from fireflies to katydids. No fewer than sixteen varieties of migratory birds, from whip-poor-will to wild geese, were to be recorded, from arrival to start of nesting, incubation, fledging of young, and autumn departure.

This was nothing unusual at the time. Sarah Mann, no matter how weary from her endless choring and life on the farm in nineteenth-century Petersham, noticed the natural world around her every day, beginning each journal entry with a description of the weather. When fresh green beans were finally on the table at her house, the event was so long anticipated and worked toward, she recorded it in her diary. Whatever was going on in nature, it mattered—a sensitivity and focus lost for many today.

To be sure, much has also been gained since the days of commonplace phenological practice. Perhaps like us, Sarah Mann would gladly have traded her daily immersion in the rhythms of nature for a cornucopia of fresh produce readily at hand year-round, in a grocery store she could easily drive to. But much has also been lost in our changing relationship with nature. For many of us now, nature is just scenery, mostly disconnected from our human-centered lives. Nature has become a place we think of as being "out there," optional entertainment, something we visit. Not a primal force in which we are acutely aware we are always and forever embedded.

Some school districts, environmental learning centers, and

ecological research stations, including the Harvard Forest, are working to help teachers and their students rebuild literacy in the language and function of the natural world, and to reawaken a connection to nature through the practice of phenology. Like an indigenous tribe trying to regain a lost language, it is an uphill battle against a thousand competing attractions indoors and on-screen. "Back when most people were still farmers, it was relevant," said O'Keefe, who helps lead the Forest's schoolyard phenology program used in more than fifty schools in the region. "Now trying to get people going out and looking again, we are starting from a much lower background," O'Keefe said. "A nature deficit disorder."

While mainstream science left phenology behind long ago, researchers are now rediscovering the value of data sets such as O'Keefe's, as well as historical records, to reveal how seasons have changed over time because of climate change. Evidence that seasonal timing has changed is apparent in old photographs, records of birding and garden clubs, even in art and in literature. The daffodil of Wordsworth and Shakespeare has advanced its bloom time so drastically as to no longer fit its literary frame: "Daffodils, / That come before the swallow dares, and take / The winds of March with beauty," Shakespeare wrote in *The Winter's Tale*. March. Not in January. And certainly not at Christmas, as happened in 2015, the *Guardian* reported, as the United Kingdom witnessed its warmest start to December in fifty years. At this rate, Britain's native daffodil, the Lent lily—named for its expected February–March bloom time—is going to need a new name.

Among the more spectacular troves of old phenology data being mined by scientists is the work of author, naturalist, and conservationist Henry David Thoreau. When he wasn't writing *Walden*, the celebrated transcendentalist was meticulously recording the first-flowering dates of hundreds of species of plants in Concord, Massachusetts, a record he compiled assiduously from 1851 to 1858. Using those records, Professor Charles Davis, director of the Harvard University Herbaria, and his collaborators reported in 2008 that nearly 30 percent of the species recorded by Thoreau had become locally extinct, and another 36 percent are barely hanging on. Hardest hit were beloved native plants—including some of the most charismatic wildflowers in New England, from orchids, asters, and buttercups to roses, dogwoods, and lilies.

The researchers built on that work to report in 2010 that the success of some plants and not others was not random, but linked to the type of plant and its ability to shift its flowering time. Invasive species that more readily adapt their flowering time to earlier springs had fared well—a finding since corroborated by more researchers. These findings have important implications for the fate of plant communities elsewhere as temperatures warm. Possible effects include expansion of species into new areas, intermingling of species that didn't used to share territory (and may not survive if they do), mismatches in timing between species that are supposed to be in synchrony, and species extinctions.

Then in 2013, Elizabeth Ellwood at Boston University, Davis, and their collaborators, examining effects of record warmth in the northeastern United States in 2010 and 2012, discovered many plants had flowered earlier than ever before recorded—as much as three weeks earlier in spring than in the historical data sets— and in some cases six weeks. A finding they could make by studying flowering dates of the same plants compiled by Thoreau in Concord in 1852, and Aldo Leopold in Wisconsin in 1935. The linkage between climate change, warming, and the biological response of plants was unmistakable.

There is even more than data, though, in these sorts of phenological records, so carefully compiled. Also to be seen is the pleasure, interest, and connection with the natural world gained by a habit of frequent, regular, simple observation. Not as a chore, but as a practice. A way of living in the world in a stance of wonder and alert, appreciative attention, with a lively, questing curiosity about the workings of our planet. Here's Leopold in his paper "Phenological Record for Sauk and Dane Counties, Wisconsin, 1935–1945":

Each year, after the midwinter blizzards, there comes a thawy night when the tinkle of dripping water is heard in the land. It brings strange stirrings, not only to creatures abed for the night, but to some who have been asleep for the winter. The hibernating skunk, curled up in his deep den, uncurls himself and ventures forth to prowl the wet world for breakfast, dragging his belly in the melting snow. His track marks one of the earliest dateable events in that cycle of beginnings and ceasings which we call a year.

From the beginnings of history, people have searched for order and meaning in these events, but only a few have discovered that keeping records enhances the pleasure of the search, and also the chance of finding order and meaning. These few are called phenologists.

How beautiful this passage is, both for its appreciation of the skunk's awakening slink after its long sleep, and the connection Leopold draws for us to the pleasure of noticing the world around us and seeking to understand the land's inner workings. He kept track of the first crowing of the pheasant, the first arrival of the marsh hawk, the emergence of the woodchuck from hibernation, and arrival of the bluebird. When the gray chipmunk first popped out of its burrow in spring, and eastern meadowlarks arrived, Leopold made a note. He tracked the first time the prairie mole made its active run up and around in broad day, the breakup of the ice on the Wisconsin River, and the arrival of the killdeer. He logged with precision the first calls of the Canada goose, woodcock, and leopard frog. He noticed when the first adult moths of the spring cankerworm flew in the trees, and the first song of the cardinal, brilliant red in the still-bare trees. He marked well the first bloom of the pasqueflower, the wood sorrel, and the toadflax, the bird-foot violet, penstemon, and coneflower. What a wonderful way to live.

Here's Ralph Waldo Emerson—on whose land Thoreau built that cabin—in his poem "May-Day":

Ah! well I mind the calendar,
Faithful through a thousand years,
Of the painted race of flowers,
Exact to days, exact to hours . . .
I know the trusty almanac
Of the punctual coming-back,
On their due days, of the birds.

I understood, though, as I came to appreciate their world-view, that something bigger than average temperatures or flowering times had shifted since Thoreau, Leopold, and Emerson made their observations and wrote their journals, papers, and poems. It started long before the big oak sprouted a century ago, and it began subtly, with the creation of electric lighting in 1880, trumping the natural cycles of light and dark. Next, we increasingly lived our lives indoors, comfortable no matter the weather or season. By the time I was growing up in the 1950s, we could, if we chose or could afford it, also eat anything, anytime, from anywhere, at the family table. Lamb for Sunday dinner in winter, shipped from New Zealand, frozen and thawed, was a treat, but not impossible, for an ordinary family such as mine. We've grown used to subverting day length, local harvest calendars and geography, and weather to live largely de-seasoned, deracinated lives.

But now we are also, through human-caused climate change, even altering the timing of the seasons themselves. The data in historical phenology records and the revival of phenological practice is confirming what the gardeners, the hikers, and the outdoorsmen and outdoorswomen of every sort already know from their own sense of a fraying natural order. Reliable patterns of nature's pageant—Emerson's trusty almanac—are slipping their chronology. And we've changed not just seasonal timing, but our relationship with nature itself. This wondrous gyre that has for millennia governed the human and natural calendar suddenly is no longer entirely independent of us; we are influencing its turn.

John Hanson Mitchell is a master of the fine art and deep pleasure of paying attention. He is the author of *Ceremonial Time, Fifteen Thousand Years on One Square Mile*, a landmark in American natural history writing. In his book, Mitchell explores one square mile of America in an ordinary bit of New England woods and pasture, his home for more than thirty years. It is one of those books in which nothing happens, and everything happens. Not long after I arrived at the Forest, I looked up Mitchell and found him at his home on a wet winter day. What a deeply nested home: the main house was full in an uncluttered way of personal treasures of every sort. But it got better, as Mitchell showed me the collection of outbuildings he has created over the years, one for a writing studio, another a cabin retreat, a third a gazebo for sitting in his carefully tended garden. Here Mitchell practices phenology in his daily observation of the seasonal progression of nature at Beaver Brook near Littleton, Massachusetts—the place he calls Scratch Flat. Watching the progression of the seasons is second nature to him: "It's how I tell time," Mitchell said. The garden is both part of how he witnesses climate change, in its day-to-day reality, and how he copes with it.

"You forge on in the face of it," Mitchell said. "This thing of a billion other possible planets, that is good news. Maybe it doesn't matter. The sun is going to burn out. In the history of the universe we are a pretty short-lived phenomenon, the human race doesn't even register. I just carry on I guess, I take the long view. The earth is going to endure." Mitchell works strictly by hand in his garden, clearing, tending, pruning, and shaping the thousands of ornamental plants on his property. "The garden is a place, but it's also a metaphor. It's a sanctuary here for me. This takes work. I like the work. That is the main existential statement." To Mitchell, people should know about everything in their little patch of the world around them. Those who don't pay attention to the wonders in their own Scratch Flat, wherever it is, well, they are missing out.

Like O'Keefe, Mitchell sees changes in the seasonal calendar: "I have watched the gardens here for thirty years, and we used to get the first frost September eighteenth on the equinox, a light frost, then we would get a killer on the tenth of October. Now the light frost is in October and the killer is in November, and that is pretty standard for the last fifteen years. We used to be a [USDA Plant Hardiness] Zone Four, now it's Five, and we are pushing Six. I can grow plants here I couldn't grow twenty years ago. I am growing a tree native to the Smoky Mountains. I should try crepe myrtle."

Earlier that same spring that I started working with O'Keefe, I had gone out on a walk at the Forest led by Harvard biologist Paul

Moorcroft for his students in a global-change biology class. The cold, wet April day, with winter still just hanging on, was perfect for climate-change jokes. But as we walked the trails and smelled the fragrance of the mineral soil, released from frost, and alive with the scent of a billion living things, we all quieted and started to see the beauty of what was around us—and its fragility.

Moorcroft fished a key from his rain pants and creaked open the door of a shed at the foot of the Environmental Monitoring Station, another of the Forest's observation towers. Within the shack, among the stacked array of monitors and instruments wired to sensors on the tower, some were continuously measuring the gas exchange of the trees with the atmosphere. While it was too early for leaves to be out, the evergreens were already responding to the longer and milder days and starting to photosynthesize. Moorcroft wanted us to see the level of carbon dioxide in the atmosphere, measured on the tower and registered in the monitor glowing in the shed. It stood at four hundred parts per million. This was the highest ever since record keeping of this measurement had begun.

The rate of change in that number, over time, had scientists particularly alarmed. Atmospheric carbon dioxide measured about 280 parts per million back when the big oak sprouted, just a century ago. Now, because of us, atmospheric carbon dioxide had soared past 400 just in the lifetime of the big oak, and it is still climbing—to the highest level in some eight hundred thousand years. No human had ever breathed this atmosphere. This seemingly harmless gas, made of one of the most common elements in the universe, was having exactly the effect on the climate and growing seasons that some scientists had long ago predicted. But it was happening far more quickly than they had ever imagined. What the pheonologists and gardeners see, scientists are scrambling to analyze and understand. Our world is already changing.

CHAPTER SEVEN

WITNESS TREE

THIS WAS THE highest I had ever climbed. I just wanted to see what it was like up here, to experience this bit of woods from the big oak's point of view. It swayed in the wind, taking me with it. Dangling from my climbing rope and harness, I was way out of my terrestrial realm. But I was enjoying the rock of the big oak's branches and was in no hurry to come down. No, not yet. Since first meeting this tree last year, I had been wanting to get here, more than eighty feet up in its canopy.

I *was* worried about the wind when the day of this scheduled climb came, forecast with gusts to twenty-five miles an hour. It was also cold, in the twenties. But there was no last-minute cancellation. So I rousted Dusty, my cat, and hurried out of bed to put the coffee on and heat up the skillet for pancakes. I slid a hunk of butter sizzling over the hot cast iron and stirred up the batter I'd set out on the counter the night before, enjoying the tangy scent of sourdough as it bubbled. I layered on the clothes. Wool long underwear. Snow pants. My big boots with the wool liner. Hand warmers to tuck in my mittens. At nine A.M. sharp, my climbing friends from MIT and Melissa LeVangie, the instructor from my first climb, arrived. This time Melissa brought her identical-twin sister, Bear, to guide all of us up into the tree.

What a presence Bear and Melissa were together; their charisma filled any room. Tall and powerful, they had the physiques, stamina, and grace of the professional climbers they trained hard to be. They shared the same house, the same profession, and a private language of tree talk practiced in hundreds of hours of professional climbing and competition at the highest levels. Beyond climbing competitively and for work, they were leaders in their profession and generous with their time. They led tree-climbing workshops for women all over the country, tree-identification walks for novices on the town common, and even helped organize and run a wood-splitting cooperative to help the town's needy with free firewood. They packed mighty appetites, loved a good meal, and climbed with confidence and joy. Not for nothing was the mascot that Melissa tucked into her climbing pack a tiny replica of Gonzo the Great, the high-spirited performance artiste of the Muppets. They were so sunny and assured, so warm and quick to laugh, anything seemed possible in their presence. In any event, there was no turning back now. We drove up to the Barn Tower, then packed all our gear into the woods. The big oak was waiting.

I'd had one practice climb on this tree before. But I had never gotten up far—just to the first rank of branches. This time we started by heaving a throw bag up into the tree to set our climbing lines. Harder than it sounds in so big a tree, it took me probably twenty tries to wing the weighted bag with a line attached up into the tree. "Commit to your throw!" Melissa kept saying, urging me time and again to use a freer swinging motion to release the throw bag high into the blueness of the sky.

With our lines finally set, it was time for liftoff. In a blur of buckling and clipping, we got into our harnesses and sat for the first time on our climbing lines, our feet swinging free a few feet off the ground. Everything on the ground—our blue tarps with gear, the understory trees I had come to know so well—was soon far below. I had a bosun's chair set in my harness and enjoyed just leaning back for the view. I could see so much better from up here, nested in the tree's branches, how the structure of its canopy opens to the sky. Each branch was positioned to gather the sun; every twig claimed its space. It had been a long winter, and some sixty feet down, the snow was brilliant white and still deep on the ground. The sky had that radiantly clear, deep-blue depth of winter, when the air is too cold to hold much moisture. No pollen was in the air either, without so much as a pussy-willow catkin yet open. The oak's branches were studded with buds still clenched tight. Another gust gave the tree a big push. I looked across the forest canopy and could see much more distinctly its complex weave, each tree growing its branches just to the tips of the next. This phenomenon has the charming name *canopy shyness*, in which trees in a forest grow the branches of their crowns out and out, reaching ever so close to one another—but stop just short of touching.

Not far from where I sat in the crown of the big oak, I looked not only into other trees, but the metal legs of the Barn Tower. This instrument tower, soaring 120 feet in the air, was originally put up by Aaron Ellison, a senior ecologist at the Harvard Forest, to create a radio relay for monitoring his Warm Ants experiment. Like any other tower at the forest, it quickly attracted other researchers' interest. Andrew Richardson, for one, installed an impressive array of equipment on the Barn Tower, including several security cameras. They watch not for vandals—but for *leaves*. Richardson had started using the cameras together with O'Keefe's records and other data to explore the effects of climate change on tree physiology, and seasonal timing. The object was to probe the forest at a variety of scales, from individual trees, to the forest, region, and biosphere.

This all got started when Richardson was at the University of New Hampshire, working with his colleagues taking measurements of the daily and seasonal rhythms of carbon dioxide exchange between the trees and the atmosphere—the breathing of the forest. He was using instruments at the top of a ninety-foot-high tower in the Bartlett Experimental Forest in the White Mountains of New Hampshire when he got a hunch he could also be measuring a lot of other things to get a better idea of how the ecosystem worked. Which, in a project meeting one day, led to a conversation with one of Richardson's collaborators. What, they wondered, about putting a camera on the tower, with the thought that, at the very least, they would get cool pictures of the forest canopy through the seasons for presentations at science talks?

They figured they would also probably be able to tell when the leaves came out, and when they fell off, which would also be useful for estimating growing-season length—key information for scientists studying how much carbon forests pack away. Within a few weeks, they installed what was then a state-of-the-art camera, beaming its images over a wireless connection to a server on campus. When the first images came in over the Internet to their computers, they were delighted that, dinky as it was, the camera was performing just as they'd hoped. Suddenly, they could monitor their remote field site from their desks. That got Richardson thinking.

The next summer, he asked a Ph.D. student whether he thought he could use computer analysis to spot the beginning of spring green-up in the images. In just days, he created a computer program that converted the red, blue, and green pixels in the camera image to numeric values. He could then count the amount of greenness in an image. Voilà: spring, pinpointed from the pixel mix. Now the team could track the development of the canopy all the way into summer, with every day's incremental growth in the leaves showing up as increasing numbers of green pixels. And come fall, the camera's pixelated signals of leaf coloring and drop were just as clear. Suddenly, big swaths of landscape could be remotely monitored for seasonal development, over the Web, from anywhere.

It was a breakthrough for scientists interested in probing how seasonal timing signaled the ecological impact of climate change. Here was the possibility of creating a whole new kind of observatory: a remote, digital observatory, with a network of cameras that could monitor the rhythm of the seasons as they transformed the land, over as large an area as the cameras could be placed, with the information streamed to a central server where the data could be shared, archived, and analyzed. Richardson dubbed it the PhenoCam network. There had never been anything like it.

In less than a year, Richardson found funding to start a small PhenoCam network to observe forest phenology across northern New England and adjacent Canada. That was in 2007. Then the National Science Foundation (NSF) in 2011 provided money that allowed the team to expand the monitoring network. Next, in 2013, NSF kicked in more money, which the team used to involve volunteers in interpreting and analyzing more than five million images streaming into a network by then grown to some 250 sites across North America, uploading images at least once an hour, seven days a week, during daylight hours. The cameras were all over the place: from instrument towers such as those in the Harvard Forest, to weather stations and building tops, from forests to tundra to Hawaiian grasslands and the desert Southwest. The PhenoCam network brought the phenological tradition of Robert Marsham, Thomas Jefferson, Henry David Thoreau, and Aldo Leopold into the digital age. What would Jefferson have given for a PhenoCam on his beloved gardens, instead of having to wait for letters from Monticello to fill him in on what was in leaf and in flower.

Even I caught the PhenoCam bug. What, I wondered, about putting a camera under the big oak tree, pointed upward to track its foliage through the year? I put the question to friends at the Knight Science Journalism program at MIT, where my idea for this book first germinated, and to my delight they bought a camera to Richardson's specs. It arrived one late autumn day in a box from California, with a hundred feet of cable. Now what? Over several weeks, I went out to the big oak with Richardson, and Emery Boose, trying to figure out where to put the camera. A soft-spoken computer whiz with an undergraduate degree in mathematics and Ph.D. in Sanskrit and Indian studies, both from Harvard University, Boose was good company. He cheerily bushwhacked through ferns and fallen leaves on a day in late fall to the spot I had finally picked, right under the canopy on the southeast side of the trunk, where the sun plays in beautiful patterns through the oak's branches. While Roland Meunier from the Woods Crew sank a post to hold the camera's mounting bracket in the ground, we set to work pushing the connectors for the Web interface hookup through holes in the camera housing. We snaked the cable through the stone wall, replaced the stones for appearances, then laid the cable right over the ground, buried under leaves, all the way to the Barn Tower shack. From there it was just a matter of hooking it up to the network with Richardson's other cameras on the Barn Tower. With his smartphone, Boose called up the Forest's wireless network and checked the image over the Internet as I pointed the camera this way and that, trying to get it just right. Suddenly, there on the Web, for all the world to see, was the big, beautiful oak, in real time.

Now the big oak had its own Witness Tree Cam, bringing its day-to-day life to the world. Hesitant as I was at first to core the tree, I seemed to have crossed quite a line. Not only numbered, tagged, and dated, the big oak was now also under constant surveillance. This is how it goes with knowledge: once you get started, it's hard to stop, and knowing what was going on in the forest, with the click of a mouse, from anywhere, was too much to resist. Suddenly I could know exactly how in the whole four-thousand-acre Harvard Forest my one tree in particular was doing, hour by hour, all day, every day. Have a look for yourself: http://harvardforest.fas.harvard.edu/webcams.

I've sat in my study in my house in Seattle, called up the Witness Tree Cam on the Harvard Forest home page, where images from all the PhenoCams at the Forest can be seen in real time, and watched, entranced, from across the country, as sunlight and shadow play over my tree. When I feel homesick for the Forest and my tree, I check the Witness Tree Cam to see if it's sunny or raining, or just to watch the tree go through its day. I sometimes log on to the PhenoCam data archive, stored on the server at the University of New Hampshire, to upload several seasons of archived images of the big oak, just to watch it leaf out all over again. The picture is sharp as could be, and the tree that beautiful. I never get tired of looking at it.

Suddenly this wild, beautiful life is a part of my life, every day, even though it is across the country. I can watch it gather snow in winter, and subtly, slowly, go from bare branches to lush leaves. I can send pictures of the tree to friends. Put it up as my screen saver and change it every day, or hour. Make the images into greeting cards. Or check in on the tree as I would a friend. It's like Facebook for nature. It's not the same as being there, but from my study some three thousand miles from the Forest, it's pretty darn good. How this sort of capacity to be in touch with a single place, or even a single wild life, instantaneously from anywhere will change people's relationship with nature—again—is yet to be seen. But we already know how it is changing science.

Now, it is possible to observe a single view, such as the big oak, or landscape over a broad area, with greater definition than ever before possible, and over any time span we care to observe, from an hour to a day, to many years. Researchers are no longer limited only to what they can see on foot, or the occasional imagery of a satellite, available only intermittently and from a great distance. And the images are viewable and shareable by anyone, anywhere, for free. Not surprisingly, Richardson and his collaborators are still figuring out what to do with so young a science and practice that taps into the old roots of phenology in such a new way.

But exciting as all this was, just like when we cored my tree for the first time, I felt a bit like an invader; a kind of tree voyeur. Who were we to be snapping this elegant life into view at the click of a mouse, without so much as getting hot, dirty, bug bitten, cold, or wet? Without even bothering to go outside at all, or for that matter, even look out a window? Where was the scent of spring, the bite of winter, the sultry summer music of crickets? What kind of knowledge was this, anyway, and could it be true knowledge at all without those dimensions? But what capacity there was, too, in tension with my doubts, to be able to see so much, and so far—even if I wasn't really "seeing" the tree in its forest at all. Or was I?

Not long after we put that camera under the oak, Richardson came out to the Forest on a bluebird autumn day with a team of researchers from his lab, and his wife, Mariah Carbone, a research scientist at the University of New Hampshire. They brought their not-yet-year-old baby, Alatna, for her first-ever trip to the Forest, and Carbone was showing her how to scoop up and toss glowing fall leaves. Meanwhile the woods rang with the squeaks of coring and the diesel purr of Bucky, as we called the bucket truck researchers were using to clip twigs from the canopy for later analysis, back at the lab.

Bearded and thatched with every-which-way hair, outside the formality of the lecture halls at Harvard, in his forest gear, Richardson always looked to me like a cross between a lumberjack and a thru-hiker. In his Carhartts and T-shirts, maybe a fleece if it was cold, he was ever eager to climb a tower, crank up Bucky, sink a corer into a trunk, or dig a hole in the dirt, seeking a root. As the sun crossed a perfect blue sky, we stopped for a lunch break on the steps at the shack by the Barn Tower, leaves sailing to ground all around us. I sat next to Richardson, eager to talk. Between bites of giant sandwiches, I wanted to air out a few things I had been wondering about. For one, what O'Keefe had said, about how scientists thought about phenology. That it wasn't considered a science at all, but "just" natural history.

"There was no experiment being done," Richardson said. "A lot of our oldest phenological observations were made by families, gentlemen-farmer types, amateur naturalists who were just interested in developing a calendar of the seasons. In what order did things happen? If birch leaves are coming out, what birds are going to show up around the same time?" Today, the big idea is to combine those same sorts of observations as evidence of seasonal timing with other data to gain understanding about how one leaf, or a tree, a forest, a region, and even the biosphere, is affected by climate change. Scaling up observations and analysis in this way from the local and particular to the regional and global was a grand challenge that would have been impossible a little more than a decade ago, but was now in sight because of digital technology and computer analytics. These also brought a whole new value to O'Keefe's years of field observations. Here was a new ability to see and verify patterns in his data and scale it up to a much bigger picture. The lab's ideas were still expanding. Margaret Kosmala, a researcher in Richardson's lab, was inviting the public to join their efforts, launching the *Season Spotter* blog at https://seasonspotter.wordpress.com and an online hub in 2015 that anyone could use to analyze images from the PhenoCams. Volunteers were invited to help build a data set to track seasonal changes in plants—when they flower, when their leaves emerge in spring, or when they turn color in autumn across North America. Richardson's lab had also recently started using a miniature drone to gather photos of the forest canopy, bringing another high-tech twist to the practice of phenology. What would those Victorian phenologists in their lace cuffs have made of Stephen Klosterman, a Ph.D. student in Richardson's lab, buzzing a drone over the forest to photograph the trees' seasonal progression?

One day I knew Klosterman was out at the Forest because I heard his drone zipping overhead. I was out with O'Keefe on his tree survey, recording the progression of color and leaf drop in the forest, tree by tree. Meanwhile, using his drone, Klosterman was surveying acres of the Forest in minutes, with the flick of a joystick. After O'Keefe and I finished up, I walked the forest road through the glittering light of a brilliant fall day, following my ear to find Klosterman. By the time I caught up with him, he had already flown a few missions and was just about to start another. The air was clear and the humidity had dropped now that the trees were ratcheting down their photosynthesis and pumping a lot less water out of their leaves. The sun was still summer-warm though, and the light shimmering through the canopy. I was in no hurry, so I decided to spend the rest of the morning watching Klosterman at work. We got talking, and it turned out we had gone to the same college and shared a love of music. Klosterman had degrees in math and trombone performance and sang in a chorus when he wasn't piloting a drone. I liked the way he opened his eyes big and wide when he was excited about an idea, and his enthusiasm was radiant. There was something beautiful, too, about the way he had brought his artistic side all the way into his work in the lab. Later that year, he created large-format photographs framed for exhibit from images he'd snapped with his drone of the autumn foliage. Tall and lanky as a heron, he was also for sure the only guy I ever saw turned out in Richardson's offices at Harvard in a linen sport jacket, pleated shorts, and loafers, with no socks. "It's summer in Cambridge, you know," he said with a wave, and headed into the lab.

For Klosterman, a drone was a useful way to explore the seasonal progression of the tree canopy. It's cheaper than hiring a plane, and efficient. On a day in spring with a lot of transition under way in the forest, it can take O'Keefe more than four hours to observe and record the leafing out of some seventy-five trees on his loop and at the Barn Tower plot, working old style, on foot, with binoculars, pencil, and paper. But with one drone flight, Klosterman could survey about ten acres of his research plots with thousands and thousands of trees at the Harvard Forest in just six or seven minutes. The drone cruises along at about eleven miles an hour, at just under four hundred feet. The drone-mounted camera also takes sharper photos of the tree canopy, from nearer proximity, than do satellite cameras.

Not that drone research is without challenges. Ironically, it takes days of mind-bogglingly difficult work to process the images the camera so quickly and effortlessly records. Then, there's the wind. An unexpected puff can send the drone off course, and even clouds are problematic: intermittent sun bedevils the exposure setting of the camera. The Woods Crew had fished Klosterman's drone out of the treetops more than once after a mishap with a wind gust. "There is always the possibility," Klosterman said, "of turning a remote-control helicopter into a remote-control hedge trimmer." This was high-tech science, with blue-tarp accessories. "Hand me those, would you?" Klosterman said to me as I watched him work, pointing to four rocks on the ground. He placed the rocks at the corners of the tarp to weigh it down, creating a tidy improvisational launchpad. He readied his joystick controls and calibrated the camera's color balance, checked its batteries, then urged me to stand back, as the drone's blades started to whir for takeoff on another mission. No doubt about it, the drone, no bigger than a place mat, was fun in the field. But that's not what attracted Klosterman to it for his research. It was the value of the pictures he could gather with it, cruising large swaths of forest to document the seasonal change of the canopy.

None of these observational methods is better than the other; they each have their uses and flaws, and they are strongest used in combination. The automated PhenoCams and Klosterman's drone flights are a bridge between conventional satellite imagery, which Richardson also uses to track the canopy, and O'Keefe's ground survey. O'Keefe's observations provided a way to verify, from the level of individual plants, just what the bigger biological picture is saying—including the understory, which was invisible from canopy-level views. The breakthrough was the power of using all these tools together to get what no one person or method could: a look at the interplay between forests, the atmosphere, and climate *at scale*. The proverbial forest, for the trees.

For Richardson, all this was the result of a long exploration. He didn't start out as a scientist; he didn't take even a single biology class while at Princeton, where he wrote his undergraduate thesis on art auctions, analyzing the prices of fifteen thousand Impressionist and Modern paintings. He was interested in which artist's works were valued most highly and came up with a model of what a picture should sell for, based on physical attributes—such as who painted it, how big it was, and whether it was signed. The thesis won academic prizes for writing about the fine arts and was the top thesis in the economics department in 1992. Richardson graduated Phi Beta Kappa with highest honors and went on, as the top student in the economics department, to enroll at MIT, where he was offered free tuition and a stipend. He began studying for his Ph.D. in economics. But things didn't go as he expected.

"I dropped out in the second semester," Andrew said. "My classes used theory to answer questions like 'How does a change in the tax rate affect consumer decisions?' And we would use this big model, and I thought, 'Okay, but people don't make decisions like that . . . I just don't believe this.' And that made it really hard for me to get excited by it." He was doing fine in his classes, but he just didn't like it. "I came back after Christmas and I immediately got sick. I didn't go to class for a week and a half. And I thought, 'I really don't care enough about this to pick it back up.'

When I realized it wasn't what I wanted to do with my life, I decided I wanted to get as far away from Cambridge as I could." So Richardson, a Canadian who had never been to the far north, headed for Dawson City—in the Yukon—to look for work.

He also got to thinking about the time he had spent with his brother before enrolling at MIT, traveling around Australia, Papua New Guinea, New Zealand, Nepal, and Pakistan. "Here are the two of us running around PNG when our combined age was forty-three. The country was so far off the beaten tourist path, and we were doing weeklong hikes through the jungle, climbing fifteen-thousand-foot mountains, and hiring local guides to take us in dugout canoes up big rivers. It was crazy, terrifying, and strange." During those hikes Richardson became fascinated with trees. "It was the beginning of this realization that all this biology I had learned [as a kid] was just memory work. I hadn't been interested because it was just boring. But being in New Guinea, it was like 'Wow, this is really cool, I wonder what is going on.'" Richardson had heard that Yale had a master's program in forestry and environmental studies. This time, he saw the program through to a Ph.D.

Richardson had joined the faculty at Harvard in 2009, and we were talking in his corner office at the Harvard University Herbaria in Cambridge. All these twists and turns later, trees still fascinated him. At the Harvard Forest, Richardson, like so many others, had found a perfect venue for experiments.

Some of what goes on at the Forest is just about creating the conditions for observations and experiments researchers haven't even thought up yet. Walk through the forest to the Hemlock Hollow, where the wood frogs first sing, and you'll start to notice yellow-paint marks on all of the tree trunks. Led by Orwig, the Harvard Forest in 2014 finished the last of the measurements to create one of the largest long-term forest monitoring plots in the world. In an eighty-six-acre swath of forest, every stem bigger around than a pencil is measured four and a half feet from the ground—including the multiple stems of shrubs—tagged, and GPS-located in what is colloquially known as the Harvard Forest megaplot. Every tree and shrub in the plot—116,000 stems—will be re-censused every five years. All that data is available for researchers anywhere in the world to use for free in their work, shared in a network of other large-scale, long-term monitoring plots managed around the globe by research centers using the same marking protocol.

The long-term nature of the observations made at the Forest are a hallmark of the work that goes on there. That shack I visited with Paul Moorcroft that dreary spring day—where I first saw for myself the measurement of four hundred parts per million in the atmosphere of carbon dioxide—is where measurements are logged from a gas analyzer on a tower put up in 1989. In that scruffy shack, in a bit of mostly third-growth woods, researchers at the Forest have compiled the longest continuous record anywhere in the world of forest-atmosphere gas exchange. These measurements are part of how we understand red oak's adaptation to the recent rapid rise in average temperatures and carbon dioxide levels.

This business of modeling the climate is still young, and computer models can't capture the complexity of the real world, from its topography to the diversity of the age and variety of trees, to differences in soil types and the intricate workings of plant physiology. Tidy scientific research papers also can't begin to capture what it's like at the emerging edges of this field of climate models, inventing the tools to make sense of how trees work and their interplay with a rapidly changing world. The gas analyzer used to tease out the carbon exchange of trees from the air was originally designed to track the maturation of corn in fields. Yet it has emerged as the best thing researchers have going right now for working in forests. Climate models also are still in their infancy, emerging as a mainstream tool since only the 1970s. Then there's the challenge of working in the field—from drone crashes to even just keeping the instruments running.

Luckily for Richardson, he's a former rock climber. So it comes naturally to him to pack all his tools and gear into a couple of backpacks and walk into the woods when it's time to replace and repair equipment the weather has worked over. I could see the Barn Tower out my kitchen window from my house at the Forest and was surprised more than once while, say, making lunch to see Richardson climbing the tower like Spider-Man to replace a pump on a gas analyzer, or to tweak the focus on a camera. He and researcher Don Aubrecht would spend hours up there on the tower, 120 feet off the ground, talking, laughing, working in their harnesses at fixing gear, with nothing around them in the thin air, not so much as a branch.

I went out to watch them on a fix-it mission one beautiful, sunny late-winter day. This time, a camera was dead on the Little Prospect Hill tower at the Forest, and Richardson had come out with researchers in his lab to repair it. We walked through the snow to get to the tower, delighted that the three feet or so that had fallen over the winter had packed down enough that we didn't need snowshoes. I led the way, enjoying the sunlight streaming through the bare trees. Too soon, we were at the tower. We dumped our gear and looked straight up at the job at hand. Richardson got ready to go to work. Such missions were nothing new to him. There were always glitches to contend with, power failures, lightning strikes, or the Internet going out. "Over the last eight years we've learned a lot about how to keep these cameras running continuously, what can go wrong, and what we need to do to prevent that from happening," Richardson said. "In the beginning, cameras would always stop working as soon as we climbed down the tower and headed home."

He shucked off his jacket, then his big snow boots, which would never fit in the tiny spaces between the crosspieces that form the tower's support. Running shoes would have to do. He took the new camera and his tools up with him in a backpack. This tower was smaller in girth than the Barn Tower, and Richardson climbed it like a ladder, hooked in by a climbing line, ascending hand over hand and stuffing his feet in the cross members of the tower. It swayed with the motion of his climb, like the big oak in the wind. But Richardson wasn't in the canopy up there; he was way above it. While he climbed, Aubrecht and Miriam Johnston from the lab opened the shed at the foot of the tower to plug into the feed from the tower with a laptop. They were testing the image from the new camera as Richardson got it running.

"Hey, Don, are you getting focus?" Richardson yelled down from the tower. He looked surprisingly comfortable up there, silhouetted against the winter sun at the tower's tapered top. Johnston stuck her head out of the plywood shack. "That's worse," she shouted back, relaying the verdict from Aubrecht as he watched the screen. So it went as Richardson tuned the focus.

"That's it!" Johnston yelled.

"Lock it!" Aubrecht boomed from deep in the shed.

The camera was live and already recording a new round of images. There wasn't much to see in winter with the canopy still dormant. But documenting precisely when the forest awakens, when at least half the trees are in full leaf, when half have colored, and when half have dropped their leaves are all key markers in the forest's seasonal year. The work keeps turning up surprises.

Trevor Keenan with Richardson published a paper in the spring of 2015 showing that the timing of spring and fall are connected, but not in the way widely supposed. Conventional wisdom—and many climate models—held that the warmer temperatures that brought on an earlier spring would also mean a later fall, and a longer and longer growing season. But Keenan and Richardson found out that the timing of autumn correlates more closely with the onset of spring than with temperature or day length. Spring's strong control on the timing of fall somewhat offsets the effect of warming. The findings do not imply a growing season of fixed length, as the relationship between spring onset and autumn senescence they discovered was not 1:1. Rather their results suggested that current models that don't include the effects of spring on autumn lead to an overprediction of the extension of the growing seasons by as much as 50 percent under future warming scenarios. "It was a eureka moment," Keenan said. Struck by the importance of their initial findings, Keenan scaled up to investigate seasonal trends on the entire East Coast. The same pattern still held true.

There are several possible explanations. "Plants know from the history of their ancestors how long their timeline is," Keenan said. "So it makes sense they would have some mechanism built into their optimum function, to have a preprogrammed senescence . . . The question is how quickly can they learn to change and detect that the environment around them has changed?" Another theory is that once trees have filled up their carbon stores, they are finished with their work for the year, even though the weather is still fine. "They have been as productive as they need to be for the year," Keenan said. "They are done."

Leaves also take a beating up there, and the earlier they unfurl, the more abuse they are subjected to and the sooner they fall apart. I knew that from leaves squirrels had chewed down on the big oak, which I'd seen on twigs on the ground, or up in the tree when I'd climbed it for a closer look. It's a combat zone up there. Almost from the minute they grow out, leaves are studded with galls, chewed and mined and shredded by bugs, lashed by the wind, and baked by the sun. Is it any wonder they are checked out by the end of a normal year and can't keep going throughout our new super-size growing seasons, weeks longer than these trees evolved for? Still, for me, the idea of seasons that lasted longer than the leaves could stay on the trees was a lot to take in. Something about it seemed unnatural—because it *is* unnatural. It's a human-caused forcing of the climate system, imposed on a natural physiological cycle with its own timing. In effect, two types of seasons exist now: the seasons that correlate with warming temperatures, and the seasons set in the timing of living things. Keenan expected that eventually the trees would catch up, using their ability to adapt to take advantage of longer growing seasons—as trees do farther south. The question is how quickly can they can adapt? For change is coming at a rapid rate.

Long-term carbon-sequestration measurements at the forest show that trees at the Harvard Forest—dominated by red oak—have been growing faster since the 1990s, as global average temperatures and carbon dioxide levels began their most rapid rise. By now, red oak is putting on more mass than any other tree species in the Forest, and faster. True, that is partly just red oak's nature. It shoots up past the competition, including red maples, shading them out and getting the lion's share of the sun. The relatively young age of the forest, still recovering from the deforestation of the nineteenth century, also makes for this strong growth. Even hundred-year-old trees such as the big oak are still booming. But the red oak's surge is also the result of climate change, manifest in warmer temperatures on average in winter, increased rainfall, and growing seasons lasting longer than at any point before the last two decades.

With the millions of microscopic openings on its leaves, called stomata (from the Greek *stoma*, for "mouth"), the Witness Tree is also speaking truth about the changing atmosphere. Tiny but mighty, stomata have been a key influence in the climate and water balance of the planet for at least four hundred million years, dated from the earliest recognizable stomata and leaf tissue found in sediment deposits. Each stoma is usually made up of two so-called guard cells, one at each end of the tiny opening, which are controlled by cell turgor. Water vapor, carbon dioxide, and oxygen all move in and out of leaves through these openings— creating a survival challenge. But Richardson and Keenan documented in another widely read published paper that at higher carbon dioxide levels, trees, including red oak at the Harvard Forest, are working more efficiently. They don't open their stomata through which they breathe as much, or as often, to take in the carbon dioxide they need. That means they can make as much and even more food while using less water. It also suggests a shift in the physiology of trees, with profound implications for everything from water cycling to climate. The trees are changing their inner workings, using less water even as they put on more growth as temperatures warm and carbon dioxide levels rise. As Oliver Morton notes in his marvelous book *Eating the Sun: How Plants Power the Planet*, the picture that begins to emerge from all this is of human fingerprints now on the most grand to the most intimate scales of our planet. From the sky and its atmosphere, to seasonal timing, and deeper still, all the way into the structure, cellular function, and photosynthetic process of individual leaves. This is the story of the big oak and, wrapped within it, our own.

I sat thinking about all that on the steps of the shack by the Barn Tower one afternoon with Richardson, after he and Aubrecht had finished another round of maintenance on the equipment. As a cleansing October wind blew, we reflected on all the changes the big oak had witnessed in its hundred or so years. It would have seen the population of New England change from being much more rural to much more urbanized, Richardson said, tipping his head back to watch the leaves pour down from the trees all around us. It would have seen carbon dioxide levels rise from about three hundred parts per million to four hundred parts per million. It would have seen a mean annual temperature rise of about a one degree Celsius—nearly two degrees Fahrenheit. The oak lived through all those changes and also endured acid rain; witnessed the development of cars; and withstood huge increases in pollution. "And the amazing thing is, it is still there, still growing a lot, thriving," Richardson said. "And that makes trees kind of cool."

True enough. Yet even in this one forest, the ability of trees to cope with all we've thrown at them is a mixed picture. Not far from where the big oak thrives, one of the most beautiful trees in the forest is slipping away. The cause is an invasive pest, stoked by climate change: the hemlock woolly adelgid. An insect no bigger than a pencil point was destroying majestic, centuries-old trees in a magical grove unlike any other in the Forest. The Hemlock Hollow.

CHAPTER EIGHT

PAST AND FUTURE FORESTS

WHILE ALL OF the Forest was a wonderland to me, the Hemlock Hollow had a special presence that drew me back, and back. For one thing, it had that great name, straight out of Winnie-the-Pooh's Hundred Acre Wood. In my imagination, I carry a map of the Harvard Forest like Pooh's, with all due apologies to A. A. Milne's illustrator, Ernest H. Shepard. I close my eyes and see the Forest's human nucleus, humming with activity in the offices at Shaler Hall, the people all tapping away at their computers as if in Owl's House, the windows aglow in the early-winter twilight with the yellow warmth of their lights. Next to it is the happy busyness of the Harvard Forest houses, the dead ringer for the rumpus of Rabbit's Friends and Relations, cozy with the talk and visiting of scholars and students from around the world, curry in one kitchen, kale in another, fiddle music drifting out one window, rebab from another. And beyond these bright lights, the secret paths lead out.

To the pasture, with its snoozing cows, and Norma, the bug-eyed Jersey, keeping watch for a treat. Up the lane, lined with sugar maples, to the sugar shack, and its curl of sweet pungent woodsmoke in spring. I see the black walnut tree out my kitchen window, elegant in its perfect symmetry, especially in winter. There's the fence around the pasture, where the chipmunks zip along the tops of the boards, and my black cat, Dusty, likes to perch. I see the steely gleam of the Barn Tower poking up from the forest and into the sky, its cameras keeping a steady watch on the big oak and its grove. Out my front door, there's the green gate leading into the big wood, where the deep-purple Concord grapes smell so good in late summer, and the poison ivy grows up the big old sugar maple, luxuriant and glossy with its itch-making urushiol. That gate leads to the sounds of frogs, too, and the curled tips of skunk cabbage that tell of spring. It leads to the shortcut I take to the big oak, not a woods road, hard packed and broad enough for driving, but a fairy trail that threads through trees close enough for me to brush the soft tips of ferns in the understory with my fingers as I walk. It leads to a mossy wooden footbridge over a brook, where the animals travel to find open water in winter, and a grove of hemlock where porcupines snack. The green gate also marks the start for O'Keefe's phenology trail. And it's the way to the Hemlock Hollow, a beating heart of the spirit of this place.

I continually explored and walked all these places, in slippers and snowshoes, hiking boots and running shoes, mud boots and snow boots. Jogging, snowshoeing, interviewing, photographing, sketching, sound recording, and note making, but most of all, just walking and looking. Sometimes I was after something specific. But very, very often I was more like Christopher Robin, just seeking each day's new encounter. Here was a researcher photographing roots under the soil with a camera snaked down a tube. Or, zip! There went a drone overhead. Lively as the Sandy Pit Where Roo Plays, intriguing as Where the Woozle Wasn't, and with artifacts of science every bit as clever as Pooh's Trap for Heffalumps, the Forest was always an adventure waiting to be discovered.

If the Harvard Forest has Heffalumps, I'm pretty sure the Hemlock Hollow is where they live. It is a chiaroscuro realm of sun filtered through deep-green boughs and velvety shadows. A vernal pool at its heart reflects the sky. It is a place, too, of small, quiet lives: red eft salamanders on their determined march, and the understory plants that thrive in deep shade. The hobblebush, with its heart-shaped leaves that turn apricot and then burgundy in the fall, and the dainty-leaved partridgeberry and wintergreen, creeping green over the ground. Striped maple reach for the sun flecks where light finds a hole in the canopy. The straplike yellow blooms of witch hazel flicker brightly, the last flowers of the year, as the hobblebush and striped maple start to turn, their yellow and orange leaves like lanterns in the dark wood. Indian cucumber lifts its delicate leaves in a perfect trefoil, topped with a cluster of wine-purple berries. O'Keefe once dug me a root of Indian cucumber on one of our walks through the hollow, and its crisp clean taste lived up to the name. Indian pipe, ghostly white, raises its curved stems in tight, nodding clusters in the Hemlock Hollow. With no chlorophyll at all, it does not mind the gloom of a hemlock grove, for Indian pipe takes its nourishment instead from mycorrhizal fungi or the roots of trees such as beech, another master of shade.

This part of the forest has a quality of light and presence that no photo seems entirely to capture. One reason is the hemlocks' feathery, soft, flexible needles thick on its branches. While other trees self-prune, dropping branches shaded beneath their canopy as they grow, hemlocks are so shade tolerant, their branches can cascade from their crown all the way down their trunk. The result is a nearly sealed canopy that allows hemlock to dominate its kingdom of shade, setting the light budget and moisture regime. Blocking all but about 1 percent of the available light from reaching the forest floor, hemlock decides which other creatures will share its regal realm and moderates the temperature range. To me it was always particularly beautiful here in winter, with the shadows cast by the big trunks, and the sunlight in visible, angel rays of beneficence between them. Wind was softened in here, too, quieted to a sound like that of a seashell held to the ear.

The first big snow of the year that I lived at the Forest came on Thanksgiving Eve. It was a spectacular dumping that started just at twilight, transforming the Forest and pasture. In the morning I strapped on snowshoes to go exploring and discovered every branch was outlined in white. As I entered the Hemlock Hollow, the tall boughs shifted with puffs of the wind, letting loose sparkling showers of snow sliding from the densely needled boughs. Each fresh cascade glittered in the light as it shushed to ground.

How I loved that winter, with more snow than even Wisnewski, of the Woods Crew, could ever remember. The silence of these snowy woods, the softness underfoot, I couldn't get enough of it, and while others—especially the Woods Crew, doing the hard work of plowing and shoveling—groaned, I secretly rooted for more. But by April, I could see the snow cover was losing ground. I wanted to enjoy the Hemlock Hollow while still in its white winter robes. So I set out on my snowshoes for what turned out to be the last time and made my way to the fallen log that crosses the vernal pool. Using my mittens, I stacked a warm wool pillow for myself and settled on the log, just to listen. Eventually I shut my eyes, the better to concentrate on the quiet swoosh of the hemlocks' boughs, as the wind stirred their tops.

The ice on the vernal pool allowed me to sit where I couldn't at any other time of year, out in the flat white disk of its frozen surface, comfortable on the fallen log that crossed it. This pool was a timekeeper. It's where I came in late fall to watch the first ice crystals form and bounce a pebble off the thickening glaze to hear the distinct sound of winter on the way. I watched it freeze, then thaw, then fill in spring with meltwater. I heard my first wood frogs here and watched their tadpoles squiggle. This is where spring's pollen was revealed in a gold-green dust on the water. The ballet of water bugs ushered in summer. The pool nearly emptied by summer's end, as the trees drew up its water and let it loose to the air. It refilled in the fall, as the trees lost their leaves and the hemlocks slowed their photosynthesis, too.

The pool centers a grove of eastern hemlock—a remarkably tough tree. It is so long-lived and shade tolerant, it can bide its time for a hundred years in the understory, waiting for a chance to exploit an opening in the canopy, then quickly grow on. Everything about hemlock is long lasting: its litter is slow to decay, piling up deep and soft on the forest floor. It can grow in almost pure stands, defining the look and function and lives of its community.

Hemlock reached its peak abundance in New England during a relatively warm and moist climate interval from eight thousand to five thousand years ago. But then, about fifty-five hundred years ago, hemlock as a species collapsed across its range in the Northeast. Scientists still don't completely understand the cause, whether disease, an insect outbreak, a climate shift, or a combination, though climate is thought to be key.

It took about a thousand years, but hemlock eventually recovered and was again abundant in New England until the time of European settlement. The tree had struggled back only to be cut down as farmers cleared fields or harvested the tree's bark for its tannin to make leather. But by the end of the nineteenth century, hemlock regained ground as the farmers and tanners and loggers left for mills and jobs in town. It seemed hemlock's time had come again. Yet now it is facing another mortal threat.

First reported in the eastern United States in Virginia in 1951, *Adelges tsugae*, the hemlock woolly adelgid, is native to East Asia. This tiny, unremarkable insect spins a woolly, waxy casing for itself in its chosen spot: the undersides of hemlock twigs at the base of the needles. Western hemlocks are resistant to adelgid native to their range. But the mortality of infested eastern hemlock, colonized by invasive adelgid from Japan, can be 100 percent. Adelgid in these eastern forests has no natural predators, and eastern hemlock shows no signs yet of developing resistance. So the insect is relentlessly killing eastern hemlock by draining their vital juices through a small sucking mouthpart.

Already millions of hemlocks have been infested, from Georgia to southwestern Maine, and in the Harvard Forest, too, the hemlocks are dying. Adelgid can kill even a big healthy hemlock in just four to ten years. The insects typically don't have wings, but instead are carried on the wind, from tree to tree. Eggs and adelgid nymphs or crawlers, as the adelgid's juvenile form is called, also hitch rides on the feet of birds and the fur of small mammals, aiding their dispersal far and wide. Adelgid can reproduce twice a year when conditions are favorable. The insect is killed by deep cold, but it has been following warming winters ever northward, and its populations are exploding. Amy Hessl and Neil Pederson report in their beautifully written paper, "A Paleoecological

Requiem for Eastern Hemlock," that adelgid has been flourishing and expanding its range since 1980 at rates from twelve and a half kilometers or eight miles per year in the Northeast to even twenty to thirty kilometers or twelve to eighteen miles per year in the southern Appalachian Mountains. From both the ravages of infestation, and landowners' cutting of threatened hemlocks preemptively, the invasion of adelgid has made for the demise of a beloved native tree species not seen on this scale in New England since the losses of American chestnut and American elm. Those calamities give us some notion of what to expect now, with eastern hemlock.

Chestnut was a marvel of the American forest, often reaching a hundred feet in height and growing in nearly pure stands, producing abundant chestnut mast. Thoreau in his journals writes of the thuds in the fall of people beating chestnut trees in the woods with sticks to gather nuts, and the rustling of, yes, thousands of passenger pigeons roosting in chestnut boughs to get in on the feast, too.

With their grand stature revealed in winter, their bloom in summer, and long, toothed leaves turned orange and yellow in autumn, chestnuts were also beautiful in every season of the year. Chestnut wood was softer than oak, but lightweight, rot resistant, plentiful, and easily split. The straight-grained, strong, durable wood was used to frame barns and homes all over New England. The Sanderson farmhouse was partly framed with chestnut. I have a fine shelf in my study at home in Seattle, milled for me by Roland Meunier of the Woods Crew as a memento from a big old chestnut beam taken out and replaced in Community House. It's beautiful wood, with a lovely golden and rosy grain.

Chestnut may have been the most important hardwood in America in the beginning of the twentieth century, writes Eric Rutkow in his book, *American Canopy*: "Americans rode on chestnut-paneled trains running along chestnut rail ties to reach jobs behind chestnut desks to receive messages transmitted over chestnut utility poles. They dined on chestnut stuffing at chestnut tables while wearing leather clothes tanned with chestnut." But all that came to an end beginning in the summer of 1904, when the chief forester of the New York Zoological Park, today the Bronx Zoo, noticed leaves on one of the park's trees had withered and turned brown. By the next year, every chestnut in the park was infected by a fungus that killed the tree by a form of girdling; the mycelium of the fungus encircled the life-giving cambium, killing the tree but for its roots—and producing uncountable wind-borne spores.

These entered the bark of other chestnut trees through the slightest abrasions, spreading the disease. By midcentury, the blight had devastated some two hundred million acres, wiping out virtually every mature chestnut along the way. All that remained were the indomitable stump sprouts, shooting up new growth each year only to die back from the blight once more. The noble American chestnut has been relegated to the understory. "The regal species that once lorded over the eastern forests had all but disappeared," Rutkow writes. "Gone from the mountains, gone from the woodlots, gone from the parks and gardens, gone from the life of the nation."

Incredibly, at the same time, entire blocks of Main Street America were being denuded as Dutch elm disease destroyed the country's favorite street, town-green, and dooryard tree. The outbreak was initially detected in Cleveland in the 1930s, as American plant pathologists were already fighting a losing battle with chestnut blight. The disease was eventually traced to elm burls for furniture making that had been infected with the fungus, and its vector was the elm bark beetle, unwittingly imported to the port of Baltimore from France. By the 1980s an estimated seventy-seven million elms were dead, many of them giants grown to vaulted cathedrals of green, embowering Elm Streets all over America. It is a loss the country has never quite gotten over; in his book *Republic of Shade*, Thomas Campanella writes movingly of the outpouring that resulted from his author's query in the *New York Times Book Review*, seeking people's memories of American elms. The result was a lamentation from elderly correspondents all over the nation.

Stately elms can still be seen today. In Athol, the nearest larger town to Petersham, I enjoyed the soaring lines of several elms that survived on its town common. With elms now so scarce, few hosts remain to sustain the beetles that are vectors for the disease, and so some elms such as these abide. Others, including the elms in the Old Yard at Harvard College, are sustained only by constant, expensive maintenance. So today we see chestnuts and elms in vastly diminished stature and presence, reminders of a bitter lesson: that trees once so common as to make their demise unthinkable can be nearly eliminated from the canopies of our forests and towns. And that was before the familiar devastating duo of insects and fungi was joined by another threat: climate change, which aids their spread to new locations and environments.

In the Hemlock Hollow, the snags of fallen chestnut lean against dying hemlock, in a sad tableau that hints of what is to come. In some New England landscapes, entire ridgetops and river valleys and forest groves are being dramatically altered. Already a wholesale changeover in the Hemlock Hollow is under way, from the amount of light reaching the forest floor, to the bird communities it shelters. These green, verdant woods are turning to skeletons of gray.

David Orwig not only knows his tree coring; he also is an expert in the effects of the adelgid at the Forest and throughout southern New England. I was heading out to the Forest for a walk one late winter day when he was coming back to Shaler Hall with a sprig of hemlock sticking out of his wool shirt pocket. He stopped to take it out, flip it over, and show it to me: the underside was thickly studded with the white, waxy mounds of adelgid. As cold as the winter had seemed to us, it wasn't cold enough to knock back the adelgid, Orwig said, with a grim shake of his head.

Orwig had been at the Harvard Forest since 1995, taking a job specifically created to start a study on the hemlock woolly adelgid when the insect was first gaining ground in Connecticut—the prelude to the destruction under way at the Harvard Forest today. He fell into the job after his graduate work at Penn State, reading an old posting on a bulletin board, and calling up to see if by some chance it was still open. "When I interviewed, I wondered, would these people be sitting around in tweed coats? I mean, it's Harvard, I had no idea. I brought a suit and the administrator said, 'If you wear that, you will not get the job.' "

As a boy in Pennsylvania, Orwig loved exploring the small stream behind the house and roving the adjacent acre or so of woods on his family's land in Pennsylvania. He was outdoors in all seasons learning how to identify trees and plants and still fondly remembers his seventh-grade leaf collection. The stream where he spent so much time, exploring and fishing, was lined with hemlock. So perhaps it was not surprising that at the Forest, Orwig would become hemlock's biographer, and even its undertaker, investigating the ecological ramifications of its collapse. In the hemlock's passing was easily a lifetime of inquiry. What happens to the Forest? To the stream flows, the animals and understory plants? What comes next?

Orwig had spent countless hours in these woods; he managed the marking of the megaplot, venturing into an icebound beaver swamp near the Hemlock Hollow in winter to help out with the work. He seemed always to be thinking about trees, even when not among them. I couldn't help wondering, as he confronted the task of compiling hemlock's obituary in the New England woods, did a professional scientist such as Orwig feel grief? One afternoon I walked down to his office to ask him. What about the Hemlock Hollow, a place we both knew and loved? Where he had worked, walked, thought, and studied amid these stately trees, many of them centuries old?

He paused before answering, "Forests are resilient. It's still going to be a forest. But not the one we appreciate, the spongy feel, the dark, the big old trees, that is something we are going to miss. For me, it is very real. I am sure we are going to lose them all, it is happening fairly quickly." He wondered if some of the hemlocks should be preserved. "I don't think we should watch it all just go, everywhere. What if our great grandparents could have protected chestnut, or elm? That is where we are at. We have the ability to protect some trees, even if we can't protect all of them." But is curating a hemlock zoo the right way to go? So far at the Harvard Forest, the decision had been to step back, let nature take its course. "I feel I have fewer and fewer answers as the years go on," Orwig said. "I guess it's why it's important to go out and enjoy what we have." It was ghostly to walk in the Hemlock Hollow now, to know what the trees could not.

Chestnuts and elms must in their day have seemed as inevitable a part of our landscape as sky or clouds. We just don't think about these trees that we consider practically family disappearing before our eyes. And so it is now with eastern hemlock. And even before its reign ends, already the upstarts are muscling in. We have seen this before. Red oak fills a similar ecological niche as American chestnut and avidly took its place in the Forest as chestnut succumbed. Today red oak is the dominant species in the Forest. As hemlock falters and struggles, and its canopy shreds, letting in light, black birch is seizing the moment. After hemlock's fall, this green grove will still be a forest. It just won't be the forest so many have come to know and love, that special cool refuge on hot days, quiet and deliciously dim, like a library with the shades drawn. And what of the refuge hemlock provides in winter for animals—for deer seeking respite from deep snows, the porcupine resting in its dense branches or snacking on its nutritious branch tips?

And in the summer, what of the vernal pool? Scientists have found that black birch and red oak, which often replace hemlock, use almost twice the water that hemlocks do. Will the vernal pool dry out too soon to support the wood frogs that depend on it today?

In the long scale of forest time, nothing lasts forever, and perhaps, just as before, in another thousand to fifteen hundred years, hemlock will be back yet again. But can you say losing hemlock now doesn't matter? I don't think you could say that to the newt. Not to the wood frog. Not even, I would say, to the people who have come to love it. This much we already know: a bit of diversity of the native New England woods, a basso profundo note of lush shade and cool refuge, will be missed.

I tell the story of hemlock because it is important to know that how forests are faring in our changing world is not one story, even in one forest. For now, red oaks are surging, thrusting their crowns above and beyond the competition, as they are wont to do, and putting on biomass bigger and better than any other tree in these woods. But there is no way to know, for one, if their reign, too, will be cut down in the future by an invasive pest, a change in rainfall or temperature. Already in this forest chestnut are stunted perhaps forever, and what seemed a triumphant return to power by hemlock has turned out to be a limited run. Towns in

Massachusetts are hacking down trees to prevent the spread of invasive Asian long-horned beetles, and the emerald ash borer, native to eastern Asia, threatens yet another native tree species—all of the straight, tall, noble varieties of North American ash. And so it goes, in forests throughout the United States, and the world. Forests are under stress as never before as global trade brings new pests and pathogens, and global climate change stokes heat and drought, leaving trees more vulnerable to fire, pests, and disease. With only one or two generations per century, trees struggle to respond fast enough to these challenges. We did none of this on purpose. But to feel nothing in the plight of some of mankind's oldest companions on this earth doesn't feel right, either.

My own thinking about the demise of hemlock was deeply affected by a visit to the paleoecology lab at Shaler Hall one winter afternoon, to see Wyatt Oswald, a professor at Emerson College in Boston, and Elaine Doughty, who worked not only in the Forest archives but in its paleoecology lab. They were working up a core of mud pulled from a New England pond. Paleoecologists deal in pollen, mud cores, and ecological change viewed over long time horizons. If you want to know when grasses first emerged in the New England landscape, say, or to test hypotheses explaining the regional die-off of hemlock thousands of years ago, these are your people. The team samples pollen—ancient grains of pollen laid down in layers and preserved for thousands of years in anoxic muds and lake sediments. Extracting and examining layers of pollen in the cores of these muds, drilled and pulled by hand, are how they look into the past.

When I dropped by the lab that afternoon, Oswald and Doughty had just taken a core from storage in the cooler and laid it out on a metal lab table to sample it. The core was some three feet long, and big around as a fire hose. Oswald started slicing the mud quickly with a cake knife, dipping the knife between cuts into a dish tub of water to clean it. Working quickly, with the fluidity of people who don't need to talk to coordinate their motions, as Oswald sliced, Doughty wrapped each dark, fudgy-looking round in plastic wrap. They would later return to these samples one by one for analysis, painstakingly processing the slices in a burner to reduce the mud to ashes and its enduring grains of pollen. In those grains was told an indelible history of this landscape. "We are still figuring out the climate history of this region," Oswald said. In the 1970s, it was thought hemlock's demise was due to some sort of biological agent that occurred simultaneously. But now investigators are asking new questions. "We are trying to get at the landscape scale and open new levels of detail about soils, firebreaks, and elevational gradients," Oswald said. "We've put forth the idea that it wasn't just an insect outbreak, it was climate, too."

About one hundred years of history was in each slice of muck less than half an inch thick. That's a lot of time laid out on the table. The mud had a bit of a rank, organic stink about it, like the mud in a swamp: the smell of decomposition and time. I was amazed to be standing amid bits of the earth's history stretching clear back to the retreat of the ice sheets, some fifteen thousand years ago. I took a look at a sample from a different core already prepared for analysis under a microscope, dialing in a view of grains of hemlock pollen at about 400x magnification. The pollen was from a sample of pond sediment dating to nearly eight thousand years ago. Yet the grains were firm, intact, glowing in the light of the microscope slide like golden windows into the past. Each pollen grain was distinct and richly textured, like a tiny moon. Evidence from a long-past world that we can't see, pollen records tell us what vegetation was irrefutably here. It is so difficult to imagine a world different from the one we inhabit today, to see a landscape that is not the one in which we live. Yet in these slices, in those grudging little grains, unrelenting in their biological integrity over thousands of years, the truth is told. Pollen taken from samples cored from the Hemlock Hollow had proved hemlock had come and gone time and again in this landscape over thousands of years. It was hard, with the hemlock still here, to believe there was a time—a long time—when hemlock as a species was not here at all. And then, hemlock came back.

History written in the land tells us that over the long march of time there has never been a constant or steady state in nature. Ecological and landscape change is constant. Species that do not survive one epoch may return in another, or others will evolve, even if our planet resets all the way back to rock and ice. Life—the most powerful force of all—will go on, with or without us. That left me thinking. Could it be that we are the most vulnerable species of all? For we have our own force to reckon with. Carbon.

CHAPTER NINE

CARBON

A SIMPLE, NATURAL element, stored carbon has its origins in forests. Not the Harvard Forest, or woodlands anything like them. But in the coal forests of our planet, dating back more than three hundred million years. These forests must have been strangely silent; there were no birds, and no mammals yet. Free of predators, scurrying supersize cockroaches more than three inches long multiplied by the millions. Big as a seagull, dragonflies whirred through the air, their two-and-a-half-foot wings churning. Giant millipedes marched over the swampy ground, with more than thirty pairs of legs rippling along their nearly five-foot-long bodies. Three-foot-long scorpions did battle, claw to claw and stinger to stinger. Towering over it all were the trees. Giant bottlebrush trees, like a modern horsetail rush, only six stories high. Or trees sprouting grassy leaves at their tops, with hundred-foot-tall, pole-like trunks, thickly armored with bark.

Long before the time of *Tyrannosaurus rex* and the rest of the dinosaurs, these forests once stretched from what today is North America to Europe, western Africa, and China. With no microbes yet on the earth that could dissolve the trees' lignin fibers, when all those trees died, they just fell over, sank into the swamps, and stayed there. Eventually, they formed peat deposits, layer upon layer, and age upon age. Under great pressure and heat, over millions of years, these peat layers were compressed and transformed. They became a black, sedimentary rock, composed almost entirely of carbon, long ago fixed by photosynthesis into those ancient trees. They became coal.

We mine and burn this coal today, all over the world. Other carbon-based fossil fuels, including oil and natural gas, were also formed from the decayed plant matter and the black ooze of diatoms and plankton sinking to the bottoms of those ancient lakes and seas, locked away in sediments millions of years ago. When we burn this fossil fuel—fossilized plants—we are bringing full circle the energy cycle that started hundreds of millions of years ago. We are taking millions of years' worth of stored carbon, locked away underground as coal, and, by burning it, instantaneously releasing it back to the atmosphere. This is the carbon in carbon dioxide. And it is the ever-increasing release of carbon dioxide into the atmosphere by us, with the coming of the industrial revolution and our burning of coal and other fossil fuels, that is the source of global climate change we are in the midst of today.

How odd that carbon, and carbon dioxide, would suddenly be such a problem for us. Carbon is the fourth most abundant element in the universe, essential to nearly all life on the earth—as well as many modern conveniences and luxuries. The graphite in pencils, the glittering diamond on your hand, your hand itself, the paper of this book—all are made or derived in whole or in part from carbon. The standing majesty of a tree is about 45 percent carbon, from carbon dioxide in the air, and locked by the process of photosynthesis into its leaves, wood, and roots.

Carbon dioxide is a chemical compound, produced either by burning carbon or organic matter or by respiration. A molecule of carbon dioxide is composed of one atom of carbon and two atoms of oxygen and is colorless, nonflammable, and nontoxic; it doesn't even smell. Comprising just 0.04 percent of the atmosphere, carbon dioxide is nonetheless an outsize, powerful force, critical to setting the thermostat of the planet. Without carbon dioxide and other heat-trapping gasses, the earth would be about thirty-three degrees Celsius, or nearly sixty degrees Fahrenheit, cooler than it is, and mostly uninhabitable to humans. It would never be above freezing even in the temperate zones of our planet, even in summer, and agriculture as we know it would be largely impossible.

Preventing this moonlike state of affairs are the greenhouse gases in the earth's atmosphere, including carbon dioxide, so-called because they function a bit like the glass in a greenhouse. As it enters our atmosphere, the radiant shortwave energy of the sun is transformed to long-wave radiation—heat. Molecules of carbon dioxide in the atmosphere absorb this heat and vibrate as they warm, creating even more heat. They also re-radiate the heat energy they absorb in random directions—including back to the earth. The effect, as these molecules block the re-radiation of heat back to space, is just the same as wrapping yourself in a blanket or putting on a jacket to reduce heat loss. The trouble we face now isn't that carbon dioxide is an alien force, or even in and of itself problematic. There is just too much of it. And more all the time.

The amount of carbon dioxide in the air before the industrial revolution was largely the result of natural processes. Carbon is exuded by weathering of rocks, respiration by plants and animals, and even volcanic eruptions. It also is consumed by photosynthesis in plants, and the mixing of carbon dioxide into the seas by the waves. This natural balancing or tuning effect governs the amount of carbon in the atmosphere, land, and sea. That in turn affects everything from the pH of the seas to global average temperatures.

But when we extract oil and coal from deep within the earth and burn it in ever-increasing quantities, we disrupt this natural carbon cycle. We are releasing the stored carbon from millions of years ago back into the atmosphere as carbon dioxide over much shorter time frames than would naturally occur—and far faster than the natural carbon cycle can absorb it. Picture a big bathtub with a small drain. We keep pouring more and more into the tub, but the drain can only remove so much. That's where we are with regard to our atmosphere. We are causing more and more carbon dioxide to build up in the atmosphere, trapping more and more of the sun's heat. And so global average temperatures are rising.

We know the carbon dioxide in the atmosphere today is there as a result of human activities because of the burning of fossil fuels. We know this because carbon dioxide created by burning fossil fuel bears the same chemical signature of carbon from plants: those ancient coal forests, released in a burning binge that started with the industrial revolution in 1750, with emissions increasing year by year. Burning coal has the most powerful effect on the climate because coal is the most carbon-dense fuel on the earth. Coal releases more carbon dioxide pollution into the atmosphere when it is burned than any other fuel—about twice as much as natural gas.

In June 2013, the NOAA Earth System Research Laboratory announced the monthly average amount of carbon dioxide in the air had increased nearly 43 percent in just the last 150 years. Today, carbon dioxide levels in the atmosphere are the highest they've been in at least the last eight hundred thousand years, and maybe far longer. Because of the well-known heat-trapping properties of atmospheric carbon dioxide, human-caused climate change is not controversial among climate scientists. It follows logically that the more greenhouse gas there is in the atmosphere, the higher the earth's average annual surface temperature will be.

The link between carbon dioxide emissions—particularly from burning coal—and climate warming has been understood by scientists for a long time. The Irish physicist John Tyndall *in the mid-1800s* was the first to understand the heat-trapping capacity of carbon dioxide and other gases found in the atmosphere. Svante Arrhenius, the Swedish physicist (1859–1927), went the furthest in computing the ratios at work in the greenhouse effect. He understood the linkage between the burning of coal and the loading of carbon dioxide into the air and calculated that if we doubled carbon dioxide in the atmosphere, it would increase surface temperatures by four degrees, and if we increased it fourfold, the temperature would rise by eight degrees.

That was in 1906, and interestingly with a pencil and paper he came to roughly the same conclusion as the Intergovernmental Panel on Climate Change in its final report of 2015. But unlike the IPCC, Arrhenius saw climate change as a good thing, writing in his book *Worlds in the Making*, "By increasing the percentage of carbonic acid in the atmosphere we may hope to enjoy ages with more equable and better climates . . . ages when the earth will bring forth much more abundant crops than at present for the benefit of rapidly propagating mankind." The difference was that Arrhenius believed climate change would unfold as it had in the past, over many thousands of years. It was Alexander Graham Bell, best known for inventing the telephone, who was the first, in 1917, to coin the term *greenhouse effect*. He also understood the danger of global warming and warned against the unchecked burning of fossil fuels.

Burning coal didn't start with the industrial revolution—that was just a change in scale. Coal produces a lot of energy for its weight and has been central to the comfort and prosperity of people around the world for thousands of years. Cavemen heated with coal. In America, Hopi Indians used coal to bake clay into pottery. European settlers found coal in the United States in 1673, and the first commercial U.S. mines began operation in Virginia in the 1740s. In their fascinating article "Hydrocarbons and the Evolution of Human Culture" in the journal *Nature*, Charles Hall and other authors note that commercial-scale burning of hydrocarbons denotes a key evolution in human affairs:

> The principal energy sources of antiquity were all derived directly from the sun: human and animal muscle power, wood, flowing water and wind. About 300 years ago, the Industrial Revolution began with stationary wind-powered and water-powered technologies, which were essentially replaced by fossil hydrocarbons: coal in the nineteenth century, oil since the twentieth century, and now, increasingly, natural gas. The global use of hydrocarbons for fuel by humans has increased nearly 800-fold since 1750.

In the United States, people began using coal and other fossil fuels to replace whale oil, beeswax, kerosene, and wood at home, and waterpower, windmills, and wood and charcoal furnaces for everything from milling grain to pumping water to smelting and forging metal. Fossil fuel powered the steam engines of the industrial revolution and lit cities and homes with gaslights. By the mid-nineteenth century, coal-powered steam engines powered the transportation revolution of railroad locomotion. By the time the big oak sprouted, more than a million Model T cars a year were coming off Henry Ford's assembly lines, and our carbon love affair was under way in earnest, as we compressed space and time even further with flight and created a national culture of car travel. Sarah Mann's oxen and horses gave way to tractors, and small-scale organic farms to agribusiness, deploying an arsenal of petrochemical-derived fertilizers and pesticides. Carbon dioxide became a peculiarly human respiration, not only from our bodies, but also from our modern ways of life. Now we and our planet are suffering from too much of a good thing.

The rate of greenhouse gas emissions has not been steady since the industrial revolution began. About half of the human-caused or anthropogenic carbon dioxide emissions between 1750 to 2011 occurred in the last forty years, with 78 percent of those emissions contributed by fossil fuel combustion. From 2005 to 2014, the average annual rate of increase in carbon dioxide in the atmosphere was 2.11 parts per million (ppm)—more than double the increase in the 1960s. As even those nineteenth-century scientists so well understood, increasing the amount of carbon dioxide in the atmosphere directly affects the earth's average surface temperature. Sure enough: the earth in 2014 saw its warmest year since record keeping began in 1880, according to two separate analyses by NASA and the National Oceanic and Atmospheric Administration (NOAA), a conclusion reinforced by others making measurements around the world. Then 2015 set a new record. The ten warmest years in the instrumental record, with the exception of 1998, have now all occurred since 2000. The thirty-year period from 1983 to 2012 was likely the warmest three decades of the last eight hundred years—and it would have been even warmer if not for the ocean, which has absorbed most of the heat. Meanwhile most experts think carbon dioxide levels in the atmosphere will go beyond doubling from current levels by 2100, and under worst-case scenarios probably triple. People debate whether we have entered the Anthropocene, a new geological epoch of our own making. But there's no disputing the changes we have made to our world in just a few centuries of fossil fuel burning are of the sort usually associated with geologic time scales.

I do think it is important, as we encounter all this, to remember that no one intended to cook the climate. We eagerly turned from wood to coal and oil to power industries and warm our homes. We invented these things to live better, more interesting, prosperous, easier, longer lives, and to expand our world and our knowledge. All along, we were inadvertently altering the atmosphere with carbon dioxide pollution, at rates that have accelerated right along with our knowledge and ingenuity. And now? We have, in the geologic timescale of a blink, disrupted multiple geochemical cycles and forces. From acidifying the seas, as carbon dioxide dissolves to carbonic acid, to warming the atmosphere, because of carbon dioxide's heat-trapping properties, our world is changing.

The earth is an intricately connected system of physical and biogeochemical cycles and interactions. The chemistry and temperature of the air, the winds, the mixing of ocean currents, the pH of the seas, the mechanics of food chains, the interactions of plants and animals and their home ranges, the fate of forests and the breathing of the trees—these are all connected. Change any part as fast as we have, and all the rest will cascade into new interactions that are already changing things for us, and many other beings with which we share the planet. *Global warming, climate change*, these are useless terms that fail to communicate what is really happening, notes Oliver Morton in his book *Eating the Sun: How Plants Power the Planet*. It isn't just that we have warmed the atmosphere. We have created an entirely *new system*, with feedbacks of its own.

What of the future? If projections in the upper end of the range of warming predicted for this century hold true, the average world temperature will increase between two and seven degrees Fahrenheit—an increase with no equal in the last fifty million years. Because it's what we know, we think the world we're used to is what the world will look like in the future. Actually, we've counted on it, building cities and villages packed with most of the world's population and much of our most valuable infrastructure right on the ocean's edge, heedless, even incurious, about the ancient beach lines now hundreds of feet under water. We look at the world we see today as the only world that is or was or will ever be. But if we look into the past, it wasn't like this, much of the time. And surely, at the rate we are going, it's nothing like it will be in the future.

So what now? Perhaps there will be a technological fix to this problem. Will our landscapes of the future include artificial trees—gigantic atmosphere scrubbers erected to cleanse the air day and night year-round of carbon dioxide and store it underground? Will we figure out how to decarbonize our energy systems and industrial processes? Break the link between prosperity and carbon? Global warming is predicted with near unanimity by scientists around the world to unleash storms, droughts, fires, pest outbreaks, floods, and species extinctions, scaling ever upward in severity according to our failure to reduce the carbon problem and stop making it worse. Those are the broad outlines of our situation, and they are pretty simple. But it's still an easy picture for some to ignore. Or worse, to confuse or contrive to debate as opinion what are incontrovertible physical facts. Interannual variability of the natural world and climate extremes stoked by global warming also make it easy to sow doubt.

My first year visiting at the Harvard Forest saw one of the snowiest, coldest winters in recent years, and the second, while I lived there, was even colder and snowier. The governor of Massachusetts called out the National Guard to shovel out the subway system's aboveground tracks, as Boston's transit system choked on snow. At the Harvard Forest, snow lay three feet deep on the ground on the first day of spring, with more in the forecast. The Woods Crew was thawing pipes as the temperature dove below zero night after night and stayed below freezing even during the day for weeks. I relished the cold and the drifted snow that simplified the landscape, burying the stone walls and reducing the palette of the forest to blue-shadowed white. I took long walks, snowshoeing and enjoying the animal tracks. Here was an otter slide to water, there the holes dug by gray and red squirrels digging out caches of acorns. I saw the foot prints of fishers—glorious, weasel-like carnivores—and even delicate marks in the snow from the wing sweeps of grouse, fluffing up in the cold. The paths made by porcupines traveling to hemlock groves for snacking were strewn with their fresh green nipped twigs. On just the right mornings when the sun was out and the snowflakes flat and crystalline, they flashed a brilliant prism of colors I could make sparkle and shine by turning my head to change the aspect of the light. The winter nights were deeply dark, with sharp and burning stars. By the spring equinox, it was clear nobody was going to even be seeing the ground in the Forest for a while. This, I thought to myself, is what scientists call the difference between weather and climate, as my car made sounds I had never heard from it before, trying to start on yet another six-degree morning.

While I froze in the Northeast, my husband at home in Seattle was cutting the grass and watching flowers burst forth in the warmest winter on record. I lived this tale of two cities, going home to sample a little spring for myself. For a few sweet sixty-degree days I was giddy with chlorophyll and early-spring bloom. Then I came back to the Forest, shoveled out the walk to my house, and struggled to get the front door open against a thickening dam of ice drooping down from the roof. Weather at the Forest continued to startle: record snow was followed by a near-record dry spell in April and May, then drenching rain fell, causing flooding in Boston. Aaron Ellison, a senior ecologist at the Harvard Forest, points out the sorts of extremes we are experiencing are exactly what we should be expecting: "Every day, it's another anomaly. This is what systems do when they reach a tipping point." Extremes of every sort are our new normal.

To be sure, changes in our planet's atmosphere, and even our climate, have come and gone before on our adventuresome earth, over its four-and-a-half-billion-year history. A wholesale reset is nothing new for our earth, whose climate is not terribly stable. Twenty thousand years ago, Seattle, Boston, and New York were deep under glacial ice. In just the last twelve thousand years, sea levels rose nearly four hundred feet, with the melting of all that ice. That's a blink of an eye, geologically speaking. And when dinosaurs walked the earth, our planet was ice-free, all the way to the poles. Fossils tell of palms, figs, magnolias, tiny primates, and horses the size of a cat in the Arctic. Crocodiles paddled around the warm waters off Greenland. Pines grew lushly in Antarctica.

Clues to our planet's tumultuous past are all around us, from the scratches of the receding glaciers' claws to giant boulders lazing in fields, made of rock from distant lands, dragged and dropped there by rivers of ice. Ancient sediments perch on mountaintops, uplifted by our restless earth, packed with the fossils from the bottom of the sea. But these changes in the earth's atmosphere, the alignments of the continents, and changeovers in whole suites of life, through five major extinctions as the planet remade itself over and over, unfolded at the pace of geologic time, over thousands and millions of years. That tempo changed with one of the planet's newest arrivals, just two hundred thousand years ago. Us.

This is the crucial way in which the global climate change under way now is different from the past. First, it is caused primarily by us, and seven billion people are trying to survive through it. And second, it is happening at a spectacularly fast rate. When global warming has happened before over the past two million years, it has taken the planet about five thousand years to warm five degrees. The predicted rate of warming for the next century though, because of increasing concentration of carbon dioxide in the atmosphere, is at least twenty times faster. The rate of increase is especially marked since the Great Acceleration, beginning in about 1950, with growing population, prosperity, and, notably, increased burning of coal for industry and to fire electric power plants, stoking emissions. In a new twist, escalating carbon emissions today are tied most strongly not to population increases, but to rising energy consumption and North American–style consumerism driven by global prosperity.

What if we could see carbon dioxide gas or smell it? What if we could see great clouds of it in the air, ballooning and billowing in the sky, getting bigger by the hour, like a summer thunderhead? What if we watched this cloud blot out more of the sky, year after year after year? Would people have a better understanding of carbon and its consequences if they watched it spew from their car, airplanes, power plants, all the daily, ordinary machines, infra-structure, and activities of modern life? That this pollution is so commonplace—we all create it every day in absolutely typical activities—is what makes it so insidious, so wound into the fabric of how we do almost everything.

The disruption of our climate system could be even bigger and faster than it is, given the amount we pollute. But the natural world is helping us: the plants, the seas, the living soil, and forests absorb carbon dioxide from the atmosphere, reducing the amount of heat energy that would otherwise be trapped in the climate system. Pastures grown back to forests have emerged as important ecological assets for capturing carbon dioxide out of the air as they consume carbon in photosynthesis. That's why in the lexicon of climate, forests are referred to as *carbon sinks*: they take carbon out of the atmosphere and store it in their tissues. Not only the stately old-growth forests and lush rain forests help. Yude Pan of the U.S. Forest Service and her collaborators discovered that the scrappy, fast-growing young forests full of intrepid trees grown up where pastures and crops once were use and store tremendous amounts of carbon. The world's established and regrowing forests soaked up the equivalent of 60 percent of cumulative fossil fuel emissions from 1990 to 2007, according to their 2011 paper. That's the lesson of the fast-growing big oak in the reestablishing woods of the Harvard Forest: the solid persistence and productivity of trees is helping to sustain our world.

The big oak had something else important to teach, too. But I had to go learn it for myself.

CHAPTER TEN

IN THIS TOGETHER

MORNING BROKE WITH a clear sky. Sunlight was already on the trees; it would be a perfect June day. I had been waiting for this chance for months. Today, we would finally climb the big oak not in the dead of winter, but in full leaf. I rushed out of bed and dressed for the climb, a little shy to break out my tree-climbing clothes, so new out of the box they were stiff. A shopping phobic, I had nonetheless noticed the tree-climbing clothes with the small green-tree logo that Melissa and Bear wore when they climbed and ordered myself an indulgent stash of Arborwear pants, shirts, and jackets. Never before had I done such a thing, but there it was, I had gone native. After starting out with such reticence, three climbs into this now I was hooked on all things arboreal, even the clothes. I put on the jacket and filled my pack with a picnic I had made the night before: deviled eggs, roast chicken, some incendiary dhal, and a dark-chocolate bar, then headed into the woods.

Over at the Barn Tower, Melissa and Bear unloaded our gear. The last time we had been together on a climb of the big oak, the snow was three feet deep. Today the ferns were hip high as we packed in. We traversed the short path and looked up to our prize: the big oak, its foliage nearly complete. The leaves were just a bit pale, and still slightly tender. They were not quite fully extended to full growth, or yet in their deep-olive-green color, with a stiff, waxy cuticle on top. But the tree's fluttering green robe was resplendent; every limb was loaded with leaves, positioned just so to catch the sun. It was an entirely different tree to encounter as a climber.

I got to work throwing a line up into the tree with a weighted bag on one end, to set my climbing line. Once, twice, three times: crash. The bag bounced off a branch and ricocheted exactly to the center of the cover on the Web camera mounted at the tree's trunk, a bank shot I couldn't have made if I'd tried. The cover shattered in multiple pieces, but the camera was fine. I put what was left of the plastic cover back on the camera, propped it level with a stick, and started throwing again.

But Melissa had brought out a special new gizmo: a giant slingshot. Standing shoulder high, it is for shooting a throw bag effortlessly into truly big trees. Melissa took the stance of a shot-putter ready to launch to give me the idea and urged me to give it a try. I dropped into a lunge position, sinking one knee into the forest duff, pulled back hard on the elastic, and let fly. *Whap* went the slingshot, and the bag sailed over the limb I'd been trying for, in an effortless arc. Encouraged, I got into a climbing harness while Bear started sending up the picnic basket and a special treat for today: a hammock, to tie up in the top of the canopy. I sent up my mobile office, too: a backpack with my computer and note-book, to draft this chapter while in the big oak. Going up seemed so easy this time, exciting and fun. I felt none of the raw animal fear from the first climb. I watched the tree change as I passed up into its different zones of life. How much rougher the bark was closest to the ground. I encountered so many lives as I climbed: hairy couplets of gypsy moth caterpillars, spiders climbing way faster than me. I reached the flare of the tree where its canopy begins and paused to enjoy the opening of its crown, where the tree divides into limbs, opening wide for big gulps of sun.

Then suddenly we were at the top, with only the tree's ceiling of green resplendent above us. And there was the hammock, too. Bear had already set it up, including a blanket folded for a pillow. I swung myself around and, settling in, could hardly believe my good fortune. We ripped into the food, then quieted, just enjoying where we were. How captivating to see a tree from within: the light was ever changing, the sun a luminous pageant of green and gold as the big oak's branches parted in the wind, letting the sun spill in, then swung back, offering light shade. Not the deep shade of the forest floor, but lacy, ever changing, with each toss of the wind. Everywhere there was movement. A black-and-yellow swallowtail butterfly cruised through the treetop, just past my shoulder; the chickadees called sweetly, alert to our presence. The leaves stirred in every direction as the wind blew, and the tree moved with it, up, down, and sideways, all at once. The hammock rode the tree in the wind, its rocking embrace amniotic and primal. I felt both at home and distinctly a visitor. I thought, what familiar and alien things trees are, all at once. They remain wild, essentially other, a kingdom apart. We need them, but they do not need us. Yet watching the oak from up here, for the first time I felt I understood clues for our own persistence. I'd noticed the oak's genius in abiding with other species above- and belowground, in a diverse, interconnected nation of lives. From the deer and the bear and the squirrel and the blue jay, to the vast, spreading nourishing mycorrhizal networks amid its roots. It seemed, I thought, rocking in the oak's embrace, that our task now is to live on this earth at least as successfully as this tree.

It felt like a lesson, a personal reckoning and ethical awakening from a human-centered or anthropocentric view, to simply grasp the reality of where we truly stand on this earth. We are not separate from nature, we are of it, and in it, and we need an ethical framework to match. We need a tree culture, a nourishing mutualism that embeds us in creation, working with one another in collaboration with nature to sustain us in our common home. From such a perspective, solutions can emerge. Without it, they likely will not. I do not mean that solutions lie only in individual action or conscience. But for governments and nations to do what they must, people everywhere must first conceive and insist on its necessity.

Really, what choice do we have? We are not exempt from the laws of nature. The planet will go on with or without us, with new suites of life emerging long after we perhaps are gone, or at least greatly diminished. Some, I know, think we are just a weedy species that has overrun its niche, that we can do nothing about climate change, and that we should just go ahead and run our little train of human needs and events hard as we can or like, until we crash. But I do not believe this.

It is truer, and more creative, I think, to recognize that global climate warming is the greatest unintended consequence in human history. That in desiring to grow beyond the world in which the big oak sprouted, we created problems we never intended. We busily built a civilization and, while we were at it, undercut the natural balancing capacities of our world. I agree that our human works are now greatly at risk—but also think that our situation is not hopeless. Ultimately, this is about relationships. With one another, with future generations, and with the other living beings with which we share the planet, now and in the future, with value all their own.

This is not a new idea, but it is one worth remembering. One of the country's first ecologists, George Perkins Marsh, said in 1864, "The world cannot afford to wait till the slow and sure progress of exact science has taught it a better economy." Better instead to use "the common observation of unschooled men; and the teachings of simple experience" to dwell as people should in our earthly home.

Here is Aldo Leopold picking up the thread again in 1948: "That land is a community is the basic concept of ecology, but that land is to be loved and respected is an extension of ethics." Leopold could have written that yesterday; it is even truer today. His call for an extension of ethics to include the natural world recognizes that we are not here alone and can no longer act like it or chart our future using such disabling, binary logic. The notion that we are separate and autonomous from nature was untrue to begin with. It is unsustainable now.

Living well in a damaged world is not only our task now, it has *always* been our task. I take heart in the journals of Thoreau. How he lamented in the 1850s at living in a time in which the big trees were already gone, and the noble animals with them. He wrote that the poem he was living was already diminished, with its finest verses cut by those who came before. "I am reminded that this my life in nature, this particular round of natural phenomena which I call a year, is lamentably incomplete. I listen to a concert in which so many parts are wanting . . . Many of those animal migrations and other phenomena by which the Indians marked the season are no longer to be observed . . . All the great trees and beasts, fishes, and fowl are gone."

But Thoreau also found wonder and beauty in the humblest seed, the geometry of snowflakes, the shape of a single scarlet oak leaf: "This leaf reminds me of some fair wild island in the ocean, whose extensive coast, alternate rounded bays with smooth strands and sharp-pointed rocky capes, mark it as fitted for the habitation of man, and destined to become a center of civilization at last . . . At sight of this leaf we are all mariners if not Vikings, buccaneers, and filibusters. Both our love of repose and our spirit of adventure are served." All that from the sight of a single leaf.

Thoreau, despite his despair at deforestation and a landscape impoverished for its lack of animals, still rooted himself deeply right where and when he lived. Both his celebration of and his grief for the natural world importantly remind us how new, and not new, our situation today is. I think also of Sarah Mann—persisting as civil war rent the country. She lost young children to fever and yearned for the peace of even one room to herself. Yet she worked, contributed, adapted, and found reasons to be grateful each and every day.

Along the way, it is also worth noticing that some things, many things, are much *better* now than they were in the times of Marsh, Thoreau, or Leopold. Even in the face of development, the big picture at the continental scale of North America is still one of reforestation on a grand scale. The regrowth of forests on former agricultural land over the six-state region of New England and beyond amounts to one of the great, accidental rewildings of our time. The results of the walkaway from hundreds of thousands of acres of pasture and farms has created some of the most densely forested regions in America. The trees aren't particularly big, or particularly old. But it is almost impossible to overstate the degree of change in New England landscapes over the past 150 years.

With the return of the trees have come the animals. During

Thoreau's time, about the biggest animal left in the woods was a muskrat. In 1850, you had to go two generations back to find someone who had even claimed to see a deer. Wild turkeys, bear, moose, coyote, bobcat, beaver, and deer all were rare or totally absent in New England in the mid-nineteenth century. Today, both by natural migration and intentional reintroduction, Thoreau's beloved "nobler" animals are back. So much so that in some places wildlife from black bear to deer are regarded as pests to people with no modern experience living in such close quarters with so many animals.

Not long after I began my year living with the big oak, I put up a wildlife camera near the tree. It was triggered by motion, snapping images day or night in any weather that would be stored on a memory card. I already knew about the chipmunks and squirrels all around, as well as some of the birds and frogs. The idea now was to see what else was there when I wasn't looking. An ecologist named Ed Faison went out with me not long before Christmas to help pick a good spot for the camera. We tried one tree and another in view of the big oak, finally settling on a middling red maple, securing the camera to the trunk at about chest height with bungee cords. We pointed the camera at the stone wall, suspecting it would be an animal highway. But Faison had warned me not to expect much in the so-called dead of winter, and I let it stay up for a couple of weeks before checking its digital memory card.

How delighted I was then and in the months to come to see the animal lives revealed in the big oak's midst. A white-tailed deer delicately stepped along the wall. A fat raccoon waddled down the middle of it, right about midnight on New Year's Eve as if on its way to an animal party. A coyote slinked up to and over the wall, its tail a long, relaxed curve. None of the animals seemed to take notice of the tree, though they traveled right under it. There was no nestling, no scratching, no burrowing or climbing. All were just passing through, the stone wall providing an easy corridor. They were alone, walking along at night, and heading in the same direction, from south to north. To see the extraordinary in the ordinary—this pageant of animals thriving in this beautiful, native woodland—after all that had happened here in the time of the big oak is to know gratitude for the resilience of nature. Here was a landscape that had been denuded of trees and bereft of its animals, now regrown to a forest and resurgent with life.

To be sure, these re-wilded woods are not the same forests European settlers first encountered. Those were forests of richness, diversity, and complexity we will not see again in our lifetimes. But this does not diminish the importance of our forests now. These fast-growing, recovering wildlands and woodlands of the Northeast are a great green wellspring of hope for the world. No matter what else the future will bring, in an uncertain world forests are a repository of only good verbs: Forests shelter. Nurture. Moderate. Cleanse. Regenerate. Provide. Connect. Sustain. They are role models of what we would do well to emulate in social and political bio mimicry, inspired by the genius of trees. Trees can be our wellspring of inspiration. More than building material, fuel, and carbon-storage utilities, forests are foundational to life on the earth, refugia for countless animals, and an endless source of human joy, renewal, and refreshment. The big oak and its forest were certainly all of that for me.

I had watched my tree through four seasons. I had climbed it with snow on its branches, its wood firm and frozen underfoot. I'd watched its first leaves emerge, then seen, as I climbed it again in summer, how its leaves had grown, each perfectly suited to its place, broad leaves in the shady reaches getting smaller and smaller as I went higher, to the airy, sunlit crown. I had seen trees change scientists' understanding of the world. And the big oak had certainly changed me. I had learned many things, but most of all this: People and trees are meant to be together, and if we work at it, that's how we will stay. Right here, dwelling in our common home on this beautiful earth, far into the future, amid the beauty and wonder of trees.

ACKNOWLEDGMENTS

I HAVE MANY people to thank for making this book possible. First among them is the crew at the Harvard Forest, beginning with David Foster, its director. David welcomed me aboard as a Bullard Fellow in forest research at the Forest, making a year's study in the woods possible. He also offered many astute and helpful corrections and amplifications to earlier versions of this manuscript. John O'Keefe welcomed me into his scholarship and weekly phenology walks at the Forest, sharing his deep knowledge and gentle, cheerful sensibility. This book would not have been possible without him. His reviews of the manuscript also helped me produce a truer and better telling. Harvard Forest senior scientists David Orwig, Aaron Ellison, Neil Pederson, and Audrey Barker Plotkin gave special assistance in guiding me through my many questions in reporting for this book. Field trips to join in their work and discussions with them about what I was learning—and seeking to understand—were invaluable joys. Research assistant Matt Lau helped me think through this story during long walks and talks, and aquatic ecologist Betsy Colburn took time from her own Bullard fellowship to explore with me the wetland and its wonders around the tree. Elaine Doughty guided me to gems in the Harvard Forest Archives, and senior investigator and information manager Emery Boose braved subzero weather time and again to keep the camera that monitors the big oak running.

Serita Frey at the University of New Hampshire and Anne Pringle of the University of Wisconsin took the time to teach me about the wonders in the soil, and Wyatt Oswald of Emerson College showed me in the Forest's paleo lab how history is written in mud. The Woods Crew and administrative staff at the Harvard Forest made every day during my Bullard fellowship better and more fun in more ways than I can count. Clarisse Hart, outreach and development manager at the Forest, helped in a thousand ways, not least of which were her macaroons and encouraging sticky notes.

The idea for this book grew from a year's immersion with the Richardson Lab at Harvard University while I was a Knight Fellow in science journalism at MIT. To all of the researchers in the lab who welcomed and helped me, and especially Associate Professor Andrew Richardson at Harvard I owe a deep debt. In particular, Andrew's careful reviews of the manuscript helped me achieve a more interesting, deeply reported, accurate book; any errors of course are my own. Associate Professor of Landscape Architecture Sonja Duempelmann's class in spring 2014, Tree Stories: Seeing the Wood for the Trees and the Trees for the Wood, at Harvard's Graduate School of Design, opened my mind to whole new ways of thinking about trees. I am grateful as well to the late Bruce Miller of the Skokomish tribe in Washington State for his generosity in sharing the teachings of his people, in particular insights about the role of trees in tribal culture and cosmology. He was the first to allow the thought in my mind of trees as teachers. I also thank the Knight Science Journalism (KSJ) program at MIT for my fellowship year, and continuing support for my work as I settled in at the Harvard Forest as a Bullard Fellow. From providing the Witness Tree webcam and even making a terrific film about the project, and many trips out to the Forest to share what I was learning, the community at KSJ maintained a sustaining and important role in this project. Climbing the big oak with Knight Fellows and staff as they made the film also brought joyful adventure to the work of reporting this book. To Melissa and Bear LeVangie, I give thanks for your friendship and showing me how to climb up into a big tree and love it. You changed my world. Charles Davis, professor of organismic and evolutionary biology at Harvard and director of the Harvard University Herbaria, welcomed me as a researcher at the herbaria. Barbara Hanrahan, who runs its front desk with élan, made it a home away from home for me. To the librarians at the herbaria, I am grateful for all your help tracking down obscure, old, and beautiful texts on the human and natural history of oak trees.

In Petersham, the Country Store fed me and provided a lively civic hub where I met and enjoyed so many friends. Robert Clark shared his deep knowledge of the landscape, and Larry Buell and Christine Mandel taught me about its history. Ellen Anderson gave me a generous welcome into the community, and Phil and Carla Rabinowitz were my home base three thousand miles from Seattle. The Athol Bird and Nature Club and Petersham Memorial Library were sources of support, inspiration, and the first audiences for this book. Allen Young introduced me to the communities of the North Quabbin as no one else could have and inspired me with his example of personal and professional courage, commitment to community, and justice. Ken Levin and Janet Palin shared their hearth and home and hearts.

Elizabeth Wales, my agent, launched all this with a book idea back in 2010 that morphed from a winter wren to a daffodil to a tree. Thank you for seeing a bigger picture in the story of phenology before I did, and bringing it to a wider audience than I could or would have. Just remember, this was all your idea.

My editors at the *Seattle Times*—Kathy Best, Jim Simon, Richard Wagoner, and Matt Kreamer—endorsed three leaves of absence to give me the time and intellectual space to do this book and welcomed me back to the wonderful creative home of our newsroom after each chunk of time away. For their steadfast encouragement I can't give enough thanks. My many friends in Petersham and at the Harvard Forest made my stay at Community House merry and bright, and I cherish memories of our times together. The team at Bloomsbury Publishing, especially my editor, Rachel Mannheimer, and art director, Patti Ratchford, brought exceptional talent to their work in bringing this project to fruition and understood a rather quirky story to its core. I am very grateful for the sharp eyes of copy editor Steve Boldt and to Kathy Belden, now executive editor at Scribner, for seeing the promise of this book at the start, and to Bloomsbury USA publishing director George Gibson for being its champion.

To my friends and family important in my life long before I met my tree, thanks for your encouragement. Finally to my husband, Doug MacDonald, whose long connections to the Quabbin Reservoir and its landscape helped spark my first visit to the Harvard Forest in September 2013, thanks for the support throughout this project and so many others. The cut flowers from our garden sent to remind me of home while I was at the Forest were particularly nice. So was the chocolate.

SELECTED BIBLIOGRAPHY

CHAPTER ONE: ME AND MY TREE

Ets, Marie Hall. *In the Forest*. New York: Viking, 1944. This children's classic set the tone for my own adventures in the Forest.

Jenkins, Jerry, Glenn Motzkin, and Kirsten Ward. *The Harvard Forest Flora: An Inventory Analysis and Ecological History*. Petersham, MA: Harvard Forest, Harvard University, 2008.

The Harvard Forest maintains an excellent and comprehensive website. For more information on the Forest and its ongoing work, go to http://harvardforest.fas.harvard.edu.

Also providing a good overview of the Forest's work, philosophy, and doings: http://harvardforest.fas.harvard.edu/annual-reports.

Visiting the Forest from Boston is easy. Overnight accommodations are available, and facilities for groups, educators, and facilitators. Solo researchers or people who just want to take a walk are also welcome. Access is free, no permit is required, and the Forest is open year-round. Stay on the trails, don't camp or bring motorized vehicles, and respect the scientific equipment you will encounter. Dogs and horses are welcome.

For overnight accommodations or other arrangements for groups, see http://harvardforest.fas.harvard.edu/visit.

Researchers, writers, artists, scientists, and scholars interested in the Bullard fellowships in forest research can learn more at http://harvardforest.fas.harvard.edu/mid-career-fellowships.

Journalists interested in the Knight fellowships in science journalism at MIT can learn more at http://ksj.mit.edu.

To see photographs of the people, places, and things I write about in this book and experience them as they happened, read the blog I published while at the Forest: http://www.lyndavmapes.com.

To see the Harvard Forest's home page on the Witness Tree project, including a film about the project and the Witness Tree webcam, go to http://harvardforest.fas.harvard.edu/witness-tree.

CHAPTER TWO: A BENEFICENT MONARCH

Campanella, Thomas. *Republic of Shade*. New Haven, CT: Yale University Press, 2003. 45–68.

Fisher, Roger. *Heart of Oak, the British Bulwark*. London: J. Johnson, 1771.

Heinrich, Bernd. *The Trees in My Forest*. New York: Ecco, 1997.

Keator, Glenn. *The Life of an Oak*. Berkeley: Heyday Books, 1998.

Kohl, Judith, and Herbert Kohl. *The View from the Oak: The Private World of Other Creatures*. New York: New Press, 1977.

Logan, William Bryant. *Oak the Frame of Civilization*. New York: W.W. Norton, 2005.

Morrison, Gordon. *Oak Tree*. Boston: Houghton Mifflin, 2000.

Mosely, Charles. *The Oak: Its Natural History, Antiquity & Folk-Lore*. London: Elliot Stock, 1910.

Pringle, Anne. "Mycorrhizal Networks." *Current Biology* 19, no. 18 (2009): R838–R839.

Trelease, William. *The American Oaks*. Memoirs of the National Academy of Sciences, vol. 20. Washington, DC: U.S. Government Printing Office, 1924.

Vogel, Steven. *The Life of a Leaf*. Chicago: University of Chicago Press, 2012.

Ward, H. Marshall. *The Oak*. New York: D. Appleton, 1892.

CHAPTER THREE: TO KNOW A TREE

Beerling, David. *The Emerald Planet: How Plants Changed Earth's History*. New York: Oxford University Press, 2007.

Holbrook, N. Michele. "Transporting Water to the Tops of Trees." *Physics Today*, January 2008. 76–77.

Thomas, Peter. *Trees: Their Natural History*. New York: Cambridge University Press, 2000.

Tyree, Melvin. "The Ascent of Water." *Nature* 423 (2003): 923.

Vitale, Alice Thomas. *Leaves in Myth, Magic, and Medicine*. New York: Steward, Tabori and Chang, 1997.

CHAPTER FOUR: A FOREST, LOST AND FOUND

Campanella, Thomas. "Mark Well the Gloom: Shedding Light on the Great Dark Day of 1780." *Environmental History* 12, no. 1 (January 2007): 35–58.

Cronon, William. *Changes in the Land: Indians, Colonists and the Ecology of New England.* New York: Hill and Wang, 1983.

Donahue, Brian. "Another Look from Sanderson's Farm: A Perspective on New England Environmental History and Conservation." *Environmental History* 12 (2007): 9–34.

Fisher Museum, Harvard Forest. All exhibits upstairs and down.

Foster, Charles H. W., ed. *Stepping Back to Look Forward: A History of the Massachusetts Forest.* Petersham, MA: Harvard Forest, 1998. 3–66.

Foster, David R. "Land Use History (1730–1990) and Vegetation Dynamics in Central New England, USA." *Journal of Ecology* 80 (1992): 753–71.

———. "The Pisgah Forest: Harvard's Living Laboratory." *Northern Woodlands Magazine,* Spring 2014.

———. *Thoreau's Country: Journey Through a Transformed Landscape.* Cambridge: Harvard University Press, 1999.

Foster, David R., and John D. Aber. *Forests in Time: The Environmental Consequences of 1,000 Years of Change in New England.* New Haven, CT: Yale University Press, 2004.

Foster, David R., and John F. O'Keefe. *New England Forests Through Time: Insights from the Harvard Forest Dioramas.* Cambridge, MA: Harvard University Press, 2000.

Raup, Hugh. "The View from John Sanderson's Farm: A Perspective for the Use of the Land." *Forest History* 10 (1966): 2–11.

Raup, Hugh M., and Reynold E. Carlson. *History of Land Use in the Harvard Forest.* Petersham, MA: Harvard Forest, 1941.

Stilgoe, John R. *Common Landscape of America, 1580 to 1845.* New Haven, CT: Yale University Press, 1982. 171–77.

Thorson, Robert M. *Stone by Stone.* New York: Walker, 2002.

Wessels, Thomas. *Reading the Forested Landscape: A Natural History of New England.* Woodstock, VT: Countryman Press, 1997.

CHAPTER FIVE: TALKATIVE TREES

Coolidge, Mabel Cook. *History of Petersham, MA.* Hudson, MA: Powell Press, 1948. 284–88.

Federal Writers Project of the Works Progress Administration, New England States. *New England Hurricane: A Factual, Pictorial Record.* Boston: Hale, Cushman & Flint, 1938.

Stand Records, Compartment III. Harvard Forest Archives. Harvard Forest, Petersham, MA.

CHAPTER SIX: THE LANGUAGE OF LEAVES

Ellwood, E. R., S. A. Temple, R. B. Primack, N. L. Bradley, and C. C. Davis. "Record-Breaking Early Flowering in the Eastern United States." *PLoS ONE* 8, no. 1 (2013): e53788.

Leopold, Aldo, and Sara Elizabeth Jones. "A Phenological Record for Sauk and Dane Counties, Wisconsin, 1935–1945." *Ecological Monographs* 17, no. 1 (1947): 81–122.

Morisette, Jeffrey. "Tracking the Rhythm of the Seasons in the Face of Global Change: Phenological Research in the 21st Century." *Frontiers in Ecology and the Environment* 7, no. 5 (2009): 253–60.

Primack, Richard. *Walden Warming*. Chicago: University of Chicago Press, 2014.

Primack, Richard, and Abraham Miller-Rushing. "Uncovering, Collecting, and Analyzing Records to Investigate the Ecological Impacts of Climate Change." *BioScience* 62, no. 2 (2012): 170–81.

Willis, C. A., B. Ruhfel, R. B. Primack, A. Miller-Rushing, and C. C. Davis. "Phylogenetic Patterns of Species Loss in Thoreau's Woods Are Driven by Climate Change." *Proceedings of the National Academy of Sciences of the United States of America* 105, no. 44 (2008): 17029–33.

Willis, Charles G., B. R. Ruhfel, R. B. Primack, A. J. Miller-Rushing, J. B. Losos, et al. "Favorable Climate Change Response Explains Non-native Species' Success in Thoreau's Woods." *PLoS ONE* 5, no. 1 (2010): e8878.

CHAPTER SEVEN: WITNESS TREE

Keenan, Trevor F., G. Bohrer, D. Dragoni, J. W. Munger, H. P. Schmid, and A. D. Richardson. "Increasing Forest Water Use Efficiency as Atmospheric Carbon Dioxide Concentrations Rise." *Nature* 499 (2013): 324–27.

Keenan, Trevor F., and Andrew D. Richardson. "The Timing of Autumn Senescence Is Affected by the Timing of Spring Phenology: Implications for Predictive Models." *Global Change Biology* 21, no. 7 (2015): 2634–41.

Morton, Oliver. *Eating the Sun: How Plants Power the Planet*. New York: Harper Perennial, 2009. 319–71

Richardson, A. D., T. A. Black, P. Ciais, N. Delbart, M. A. Friedl, N. Gobron, D. Y. Hollinger, et al. "Influence of Spring and Autumn Phenological Transitions on Forest Ecosystem Productivity." *Philosophical Transactions of the Royal Society*, Series B, 365 (2010): 3227–46.

To learn much more about the Richardson Lab and its work see http://richardsonlab.fas.harvard.edu.

To explore the world of online phenology as a citizen scientist see http://seasonspotter.org and https://www.facebook. com/seasonspotter/ and https://seasonspotter.wordpress.com. See also http://budburst.org and https://www.usanpn.org/ natures_notebook.

Many nature and environmental centers and conservation groups also offer field-based phenology training and programs.

CHAPTER EIGHT: PAST AND FUTURE FORESTS

Allen, Craig D, Alison Macalady, Neil Cobb, et al. "A Global Overview of Drought and Heat-Induced Tree Mortality Reveals Emerging Climate Change Risks for Forests." *Forest Ecology and Management* 259 (2009): 660–84.

Foster, David R., ed. *Hemlock: A Forest Giant on the Edge.* New Haven, CT: Yale University Press, 2014.

Hessl, Amy, and Neil Pederson. "Hemlock Legacy Project (HeLP): A Paleoecological Requiem for Eastern Hemlock." *Progress in Physical Geography* 37, no. 1 (2012): 114–29.

Kizlinski, Matthew L., David A. Orwig, Richard C. Cobb, and David R. Foster. "Direct and Indirect Ecosystem Consequences of an Invasive Pest on Forests Dominated by Eastern Hemlock." *Journal of Biogeography* 29 (2002): 1489–1503.

Orwig, David, David R. Foster, and David L. Mausel. "Landscape Patterns of Hemlock Decline in New England due to the Introduced Hemlock Woolly Adelgid." *Journal of Biogeography* 29 (2002): 1475–87.

Rutkow, Eric. *American Canopy: Trees, Forests, and the Making of a Nation.* New York: Scribner, 2012.

Tingley, Morgan W., David A. Orwig, Rebecca Field, and Glenn Motzkin. "Avian Response to Removal of a Forest Dominant: Consequences of Hemlock Woolly Adelgid Infestations." *Journal of Biogeography* 29 (2002): 1505–16.

CHAPTER NINE: CARBON

Beerling, David. *The Emerald Planet: How Plants Changed Earth's History.* New York: Oxford University Press, 2007.

Emanuel, Kerry. *What We Know About Climate Change.* Cambridge, MA: MIT Press, 2012.

Morton, Oliver. *Eating the Sun: How Plants Power the Planet.* New York: Harper Perennial, 2009.

Pan, Yude, et al. "A Large and Persistent Carbon Sink in the World's Forests." *Science* 333 (2011): 988–93.

Roston, Eric. *The Carbon Age: How Life's Core Element Has Become Civilization's Greatest Threat.* New York: Walker, 2008.

Stocker, T. F., D. Qin, G. K. Plattner, S. K. Allen, A. Nauels, Y. Xia, V. Bex, and P. M. Midgley, eds. *Summary for Policymakers* in *Climate Change 2013: The Physical Science Basis.* IPCC, 2013.

Weart, Spencer. *The Discovery of Global Warming.* Cambridge, MA: Harvard University Press, 2008.

CHAPTER TEN: IN THIS TOGETHER

Foster, David R., Glenn Motzkin, Debra Bernardos, James Cardoza. "Wildlife Dynamics in the Changing New England Landscape." *Journal of Biogeography* 29 (2002): 1337–57.

Fowles, John. *The Tree.* New York: Ecco, 2010.

Harrison, Robert Pogue. *Forests: The Shadow of Civilization.* Chicago: University of Chicago Press, 1992. 1–58.

Jones, Owain, and Paul Cloke. "Grounding Ethical Mindfulness for/in Nature: Trees in their Places." *Place & Environment: A Journal of Philosophy & Geography* 6, no. 3 (2003): 195–213.

———. *Tree Cultures: The Place of Trees and Trees in Their Place.* New York: Berg, 2002.

Leopold, Aldo. *The Sand County Almanac.* New York: Oxford University Press, 1948. 217–41.

Miller, Bruce. *The Teachings of the Tree People.* https://vimeo.com/5545153.

Mitchell, John Hanson. *Ceremonial Time: Fifteen Thousand Years on One Square Mile.* New York: Warner Books, 1985.

A NOTE ON THE AUTHOR

LYNDA V. MAPES is the environmental reporter for the *Seattle Times*. She researched and wrote *Witness Tree* while a Knight Fellow in science journalism at MIT and a Bullard Fellow in forest research in residence with her oak at the Harvard Forest. This is her fourth book. She lives in Seattle.

www.lyndavmapes.com